## Why Caravans Roll Over
### and how to prevent it
Collyn Rivers
RVBooks.com.au (2019)

Caravans and their tow vehicles rarely jack-knife and roll over – but when they do the results are catastrophic. The cause, and how to prevent it, explained in plain English.

# Contents

| | |
|---|---|
| INTRODUCTION | 1 |
| PART ONE - Why Caravans Roll Over | 7 |
|     CHAPTER ONE - Why and How Caravans Sway | 8 |
|     CHAPTER TWO - About Tyres | 12 |
|     CHAPTER THREE - About Steering | 18 |
|     CHAPTER FOUR - About Instability | 22 |
|     PART ONE SUMMARY - Tow Vehicle and Caravan Behaviour | 45 |
| PART TWO - Rating Your Rig | 48 |
| PART THREE - A More Technical Explanation | 53 |
|     CHAPTER FIVE - Vulnerabilities | 54 |
|     CHAPTER SIX - The Pendulum Problem | 63 |
|     CHAPTER SEVEN - The Stable Caravan | 67 |
| References | 78 |
| Acknowledgement | 84 |
| Dedication | 85 |
| About the Author | 86 |
| Publishing Information | 88 |
| Detailed Table of Contents | 89 |

# INTRODUCTION

Caravans and their tow-vehicles have a unique type of accident (jackknifing and rolling over). The cause is known and long understood. The rig is triggered from its intended stabilising characteristic (understeer) into seriously escalating unstabilising oversteer.

In conjunction with caravan and other issues, your tow vehicle's margin of understeer ultimately determines your rig's stability and especially how it will act in an emergency swerve at high speed.

If understeer is reduced to zero (or minus) the tow vehicle will oversteer. If not corrected in time the rig may jackknife, especially at high speed.

## Understeer/oversteer

Whilst cornering too fast an understeering vehicle takes up a wider radius. If hit by a side wind gust, it automatically turns slightly away from that wind. Understeer also ensures a vehicle will stay in a straight line on a cambered road. Excess understeer is rare, but can happen if the front springs of over-cab bed motorhomes are beefed up.

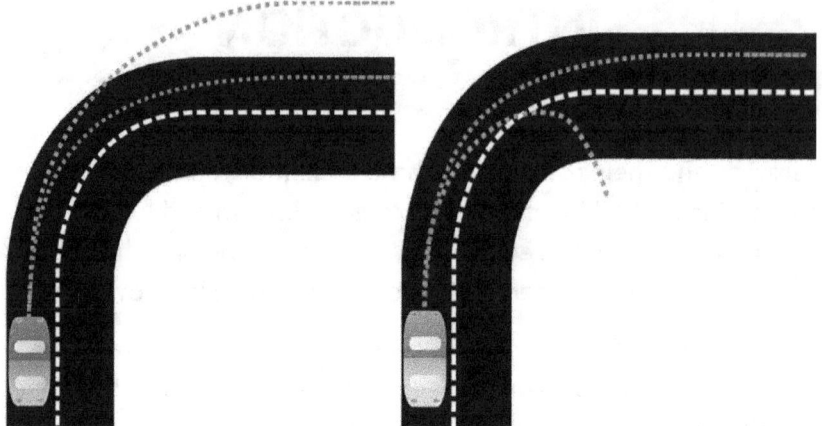

*Figure 01.* **Left:** *Green line shows desired mild understeer, red line shows excess understeer.*
**Right:** *Green line shows low speed cornering, red line effect of oversteer. White line is road centre-line marking.*
*Pic: wikipedia.org*

An oversteering car or 4WD cornering too fast behaves as shown in Figure 01 (right). If hit by a side wind it automatically tightens turning. A solo car or 4WD in an oversteering situation (as it begins to turn in) can be remedied by an experienced driver applying an appropriate amount of opposite steering lock. If that is not done soon enough, (or too strongly), oversteer is likely to escalate and the vehicle to spin.

Such correction cannot reliably be done whilst towing a caravan. It can only be corrected by centralising the steering wheel and manually applying the trailer's brakes (*not* the tow vehicle's). Any attempt to correct the rig, as for a solo car or 4WD, is likely to worsen the situation.

## Oversteer whilst caravan towing

Oversteer is the fundamental undoing of caravans towed at excess speed. If the rig starts snaking strongly, unless the tow vehicle has sufficient margin of stabilising *understeer*, that vehicle is likely to be triggered into non-correctable *oversteer* and potential jackknifing.

The necessary margin of understeer is designed in by the vehicle maker but, as this book later explains, it is degraded by excess tow ball mass, overloading (particularly at the tow vehicle and caravan's rear), or lost by incorrect front/rear tow vehicle tyre pressure ratio - Chapter Two.

Race car and rally drivers tend to prefer a margin of oversteer and know how and when to exploit it. Suspension engineers and those who understand this issue and its effect on vehicle handling can detect incipient oversteer. For the majority lacking such experience, a clue is that a potentially oversteering vehicle's steering feels 'lively', responding 'eagerly' to the steering in a sports car like manner. An understeering car's steering is less responsive, unfortunately fooling some caravan towing drivers as being 'less desirable'.

## Misleading information

Much written about caravan and tow vehicle stability discusses reducing yaw and snaking (i.e. ongoing yaw) during *normal* driving. Whilst doing that is necessary, the result may mask inherent instability. The rig may *feel* stable but still jackknife in emergency situations.

*Figure 0.2. Typical caravan jacknife. Pic: greynomads forum.*

Such situations are triggered (usually at high speed) by a driver suddenly swerving, by sudden changes in road camber and by strong side-wind gusts, including those from trucks passing in the opposite direction, and when overtaking or being overtaken by such vehicles.

This is particularly likely if towing a long, end-heavy caravan by a lighter vehicle. It may *feel* ultra-stable, but that 'stability' can preclude a required strong (emergency) change of direction. Or if it does begin to change direction, it may then not be possible to regain steerability. In essence the (caravan) tail wags the (tow vehicle) dog.

Police officers and tow truck drivers report that 'my rig always seemed so stable until then' is, the most common of all post-jackknifing reactions.

The major concern is thus not how a conventional caravan and tow vehicle behaves in normal driving. It is how it will behave in abnormal emergency situations. These include:

* Applying the tow vehicle's brakes (only) whilst descending a steep winding hill at speed,

* Swerving at speed, particularly when overtaking or being overtaken by a long truck or semi-trailer,

* Towing a long and heavy caravan at high speed by a vehicle that is lighter than that caravan (https://www.rvbooks.com.au/blog/post/4443/Overweight-RVs-a-police-point-of-view). Published police findings show that most Australian RVs are overweight in one more areas - some grossly so.

The reasons why tow vehicles and caravans go out of control are many and complex, but knowing why and how assists to prevent this.

## Critical speed

All caravans towed via an overhung hitch become unstable beyond a specific speed. That speed called its *critical speed*, and is unique to each rig and its loading. There is rarely only a single issue that causes this. There are usually a number of issues that, collectively, contribute. As many as possible need to be addressed. Your aim is to ensure your rig's critical speed is way above 100 km/h - and to *never* tow above that speed (or your country's towing speed limits).

It is currently not feasible to measure a rig's critical speed (except by wrecking it in the process) but, by following the suggestions in this book, your rig's critical speed should be raisable high enough to provide an adequate margin of stability. These changes will also reduce snaking.

## Major areas of concern

The main factors that determine stability and critical speed include:

* Mass of the laden caravan relative to the mass of the laden tow vehicle (the lower the better),

* Tow vehicle wheelbase (the longer the better),

* Tow hitch overhang (the shorter the better),

* The caravan's axle/s location (further toward its rear the better),

* Having the caravan's mass (and subsequent loading) as centralised as possible,

* Centre of gravity of the laden caravan (the lower the better),

* Sidewall stiffness of tow vehicle tyres (the stiffer the better).

This book explains how and why caravans and their tow vehicle can seem normally stable yet can, within a few seconds, jackknife out of driver control. It shows, in detail, how to avoid or correct this.

## Warning

Many advise to correct snaking by accelerating. This was fine at low speed, but if done above about 80 km/h risks exceeding critical speed. It is better not to do this at *any* speed as it becomes an instinctive action.

## Note:

As the subject is complex, this book has some deliberate repetition.

# Part One

# CHAPTER ONE

## Why and how caravans sway

There are two main types of vehicle-drawn trailers: dog-trailers and pig-trailers: (why such trailers are so-called is unknown).

## Dog-trailers

Dog-trailers have two or more axles. One axle (at the front) pivots, or has so-called Ackerman steered wheels (as used at the front of cars and trucks). Either way enables the trailer to be steered by the vehicle's drawbar.

Such trailers have long since proved stable. Many load-carrying commercial versions are used to this day – but the dog trailer configuration (Figure 1.1) is now rarely used in caravan form.

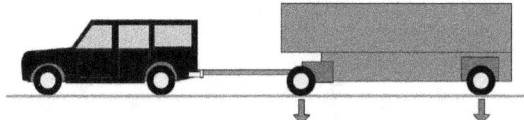

*Figure 1.1. Dog trailer. Pic: original source unknown.*

In practice, dog trailers shimmy slightly on rough roads but are not prone to 'wagging the tail' of whatever tows them.

## Pig-trailers

Pig-trailers (Figure 1.2) have single or double (and sometimes triple) axles. These are ideally positioned and laden such that the trailer is about 10% nose heavy. Chapter Four explains why.

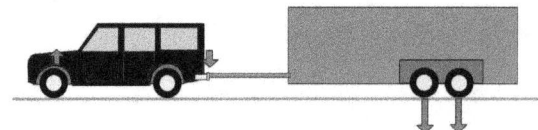

*Figure 1.2. Pig trailer. Pic: original source unknown.*

Pig trailers are coupled to the tow vehicle via a hitch at the rear of that vehicle - well behind the centre-line of the tow-vehicle's rear axle.

In their original (late 1890s) commercial load-carrying form, pig trailers rarely exceeded walking speed and stability was not a problem. By 1920, however, vehicle speeds had increased and such rigs increasingly jackknifed and overturned. Even back then, the pig trailer concept was proving fundamentally unstable.

## Hitch overhang

Around 1921, US trailer maker Fruehauf realised that if the trailer yawed in one direction, that overhang hitch not just *allowed*, but *caused*, the tow vehicle to yaw in the other direction.

If the tow vehicle or the pig-trailer yaws (sways) each, via that overhung hitch, exerts side forces on the other – but in the opposite direction. Friction and other damping reduces yaw but only effective at low to medium speeds. By 100 km/h it has negligible effect (see Chapter Four).

Fruehauf further established that the shorter the hitch overhang and the longer the tow-vehicle's wheelbase (distance between its front and rear wheels) the better. The company also found, and it later became clear why, that the laden tow vehicle needed to be at least as heavy as the laden trailer, and that the greater that ratio, the better.

## Eliminating hitch overhang

Having established the major cause of instability was hitch overhang, Fruehauf took that finding to the extreme: that locating the tow-hitch directly over the tow-vehicle's rear axle/s eliminated that instability. So why have any at all? Fruehauf's solution firmly established the fifth-wheel concept. It had been used, in 1917, in a one-0ff fifth-wheel caravan designed by Glen Curtiss, but did not become popular (mainly in the USA) until the early 1930s.

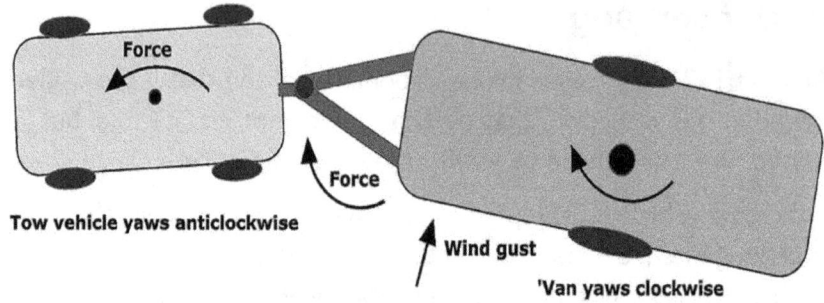

*Figure 1.3. Pig trailer yawing. Pic: rvbooks.com.au*

## Fifth-wheeler - zero yaw

If the tow hitch is directly over the tow-vehicle's rear axle neither tow vehicle nor trailer yaw will affect the other (Figure 1.4 ).

Like the pendulum of an old-fashioned grandfather clock, that trailer is free to move. Wind gusts may cause the trailer to yaw slightly, but that yaw has no effect on the tow vehicle - or vice versa. Driving a fifth-wheel caravan rig feels much like driving a motorhome.

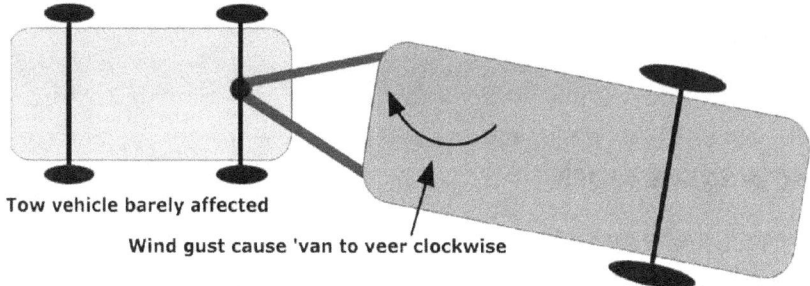

*Figure 1.4. Locating the tow hitch directly above the centre-line of the tow-vehicle ensures that if either part yaws, no forces are transfered to the other. The configuration is fundamentally stable. Pic: rvbooks.com.au.*

Whilst conventional caravan stability issues are primarily caused by tow hitch overhang, many other aspects are involved. Almost all, however, are associated with, and ultimately determined by hand-sized patches of rubber: i.e. the tow vehicle's and (but less so) the caravan's tyre contact patches that grip the road.

How contact patches do that profoundly affects every aspect of how tyred road vehicles behave. This is covered in the following Chapter.

How tyres profoundly control (stabilising) understeer, and (unstabilising) oversteer, is covered in Chapter Three.

# CHAPTER TWO

## How tyres work

In the era of iron and the later solid rubber tyres, a steered wheel caused the associated tyre and vehicle to follow the same path. If cornered too hard the tyre suddenly lost all grip and slid.

In 1888, Scottish doctor, John Boyd Dunlop, invented pneumatic tyres. Initially of bicycle proportions, they gripped better and provided a softer ride, but otherwise behaved much as described above.

Pneumatic tyres, however, became progressively made with a wider cross-section and their behaviour changed. They no longer reflected solid rubber tyres' 'either grip or not' but had a range in-between where the tyre distorted before all grip was lost. This is even more so with the radial ply tyres of today, that have some latitude between gripping or not.

## How tyres are *actually* steered

Weight on a tyre causes that part of its tread on the road to form an elongated oval footprint - called its contact patch. Those of tow vehicles and caravans are the size of an average human hand. Their grip is partly frictional and partly molecular. That grip does not follow the normal (linear) laws of friction, i.e. increasing the imposed weight does not increase grip proportionately - but by about 80%.

*Figure 2.1. Early pneumatic tyres were of bicycle size.
Pic: michelin.com*

## Slip angles

When you turn the steering wheel, the turning force acts on the steered wheel's rims and the base of their tyre's side walls. This does not directly turn the footprint: it diagonally distorts its 'contact patch'. This, in turn, causes the steered moving vehicle to follow a radius that is less than the steered wheel rims attempt to impose.

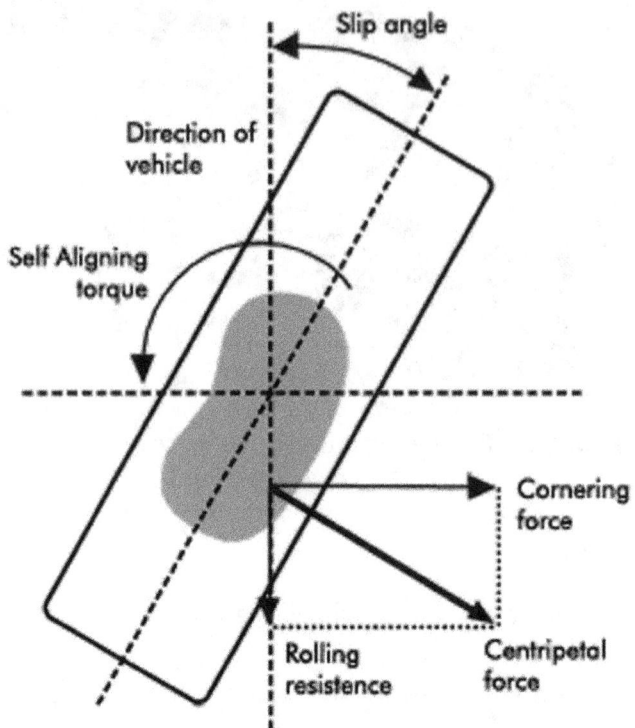

*Figure 2.2. The slip angle concept. Pic: original source unknown.*

That angular difference is (misleadingly) known as the 'slip angle'. The contact patch does not 'slip' as such: that slip angle is the difference between where a tyre's footprint (contact patch) is *pointing* and where that vehicle is actually *heading* - due to tyre side wall flexing and its contact patch (footprint) distorting.

There is a limit to that grip but on a dry and grippy surface the slip angle of radial-ply tyres is typically eight or so degrees before all grip is lost. (Cross-ply tyres had less final grip - but lost it more progressively).

Not only steered tyres act like this. The side force exerted (say) on a tow-vehicle's rear tyres by a sudden gust of wind from a passing truck generates a slip angle on its rear tyres. This literally causes those tyres also to steer that vehicle suddenly and strongly from its rear. If extreme, this rear-wheel steering action may cause a jackknife.

Major slip angles are imposed on a twin-axle caravan when tightly turned. The tyres, however, rarely slide: their contact patches distort instead.

When a caravan is attached, the essential tow ball mass (Chapter Four) pushes down on the tow vehicle's rear tyres, thus increasing their slip angle. That tow ball mass also causes the front of the tow vehicle to rise. This reduces the weight (down force) on its front tyres, thus decreasing their slip angle, undesirably reducing understeer.

## Tyre wall stiffness and slip angle

This is more of an issue for tow vehicles, but relates to a caravan's stability as well. The stiffer a tyre wall and the wider its cross-section, the better its ability to keep the contact patch stable. The best choice is so-called T-rated tyres (they have other designations, but tyre dealers will know what T-rated means). These slightly harden the ride, but offset by being designed to carry their full specified load at all times.

## Tyre pressure and slip angle

Tyre slip angle is profoundly affected by tyre pressure. Not knowing this causes many a caravan owner to prejudice stability by using totally inappropriate tyre pressures for their tow vehicle.

*Increasing* tyre pressure or reducing their loading *reduces* slip angle.

*Reducing* tyre pressure or increasing their loading *increases* slip angle.

Because of this relationship, *increasing* tow vehicle *rear* tyre pressures whilst towing usefully *reduces* their slip angle. It assists to restore their resistance to yaw forces.

*Reducing* the tow vehicle's front tyre pressures whilst towing *assists to restore* the car maker's intended (non-towing) front/rear slip angle ratio.

By and large (and within tyre maker limits), tow vehicles need their rear tyres 50 to 70 kPa (7-10 psi) higher than the front whilst towing.

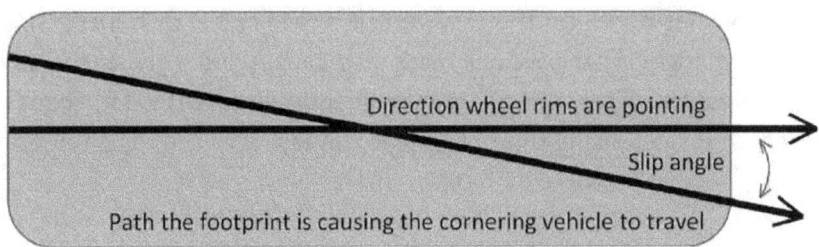

*Figure 2.4. The effect of slip angle. The magnitude of that angle is governed by the factors described in the main text.*

The pressure must be within tyre makers allowed limits, but it is that *ratio* that truly matters. Caravan's tyres are best run at the tyre maker's recommended pressure for the known weight on the tyre. If you do not know that weight, measure it (preferably on a certified weighbridge) and follow maker's tyre chart pressures. It is important not to *under*-inflate.

For dual-axle caravans a (minor) increase in stability is achieved by running the caravan's rear tyres 50-70 kPa (7-10 psi) higher than the front. Most owners will not feel any difference but a rig's margin of stability is typically increased by optimising many small adversely contributing factors as well as a few major such. Even small improvements assist.

## Laden tow-vehicle weight and caravan laden weight

This is the classic issue of the tail wagging the dog. That action is akin to that of a hammer thrower. A strongly whirling athlete can barely control the whirling 7.26 kg (16 lb) ball just prior to its release.

When released, centrifugal force can hurl that ball, whilst airborne, for eighty plus metres. (It rolls yet further after landing).

A strongly yawing (say) kg conventional caravan exerts a not dissimilar 'tail wagging dog' type of force on its tow vehicle (and vice versa) just prior to jackknifing. The tow-hitch and safety chain/s usually prevent the caravan being 'thrown' but the result is invariably violent.

*Figure 2.5. Olympic hammer thrower. Pic: Reuters Media.*

The energy suddenly released over that second or two of jackknifing is that acquired over a probable minute or more to accelerate the mass of that rig (from rest) to the speed at which that a jackknife occurs.

## Tow vehicle weight matters

There is a strong, known and totally proven relationship between stability, the ratio of tow vehicle and caravan weight, and speed.

If all else is equal, the *heavier* the laden caravan, relative to the laden tow vehicle, the *lower* the speed at which it may suddenly become terminally unstable if subject to a side force (including sway build-up).

Chapter Three shows why and how the tow vehicle's tyre slip angles are so important. It also shows why tyre pressures may need be varied for optimal towing. The caravan tyres' slip angles matter too - but not to the same extent.

# CHAPTER THREE

## Understeer & oversteer

Understeer and oversteer ultimately determine your rig's stability, and especially how it behaves during an emergency swerve at high speed. Knowing more about this greatly assists an overall understanding of tow vehicle and caravan stability.

## Understeer

If cornering or swerving too fast, an understeering vehicle automatically runs slightly wide, i.e. it takes up a slightly wider radius and that usefully reduces the cornering forces. Also, if hit by a side wind gust, an understeering vehicle turns slightly *away* from that wind. Understeer also ensures it will stay straight on steeply cambered roads.

## Oversteer

Oversteer is the opposite of understeer. An oversteering car or 4WD automatically *tightens* any turn, and even more so if speed is increased. It is correctable (within limits) by applying opposite steering lock, but if not done soon enough it will spin out of control. Owners of early (tail heavy) VW Beetles and Porches will be only too aware of this.

Oversteer is the fundamental undoing of caravans towed at excess speed. If, for example, hit by a sudden wind gust, an oversteering tow vehicle turns *into* that wind. That wind, however, will push the caravan's nose away - in the opposite direction: virtually a recipe for snaking.

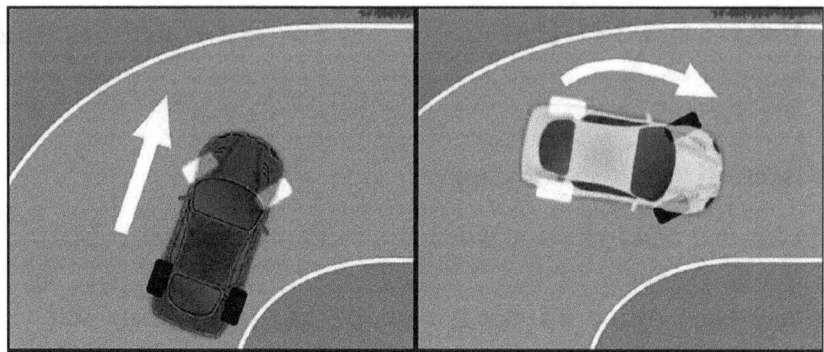

*Figure 3.1. Left: strongly understeering car. Right: strongly oversteering car. Pic: Original source unknown.*

Understeer is designed in by the suspension engineer. It can be done in several ways but primarily by ensuring the vehicle's front tyres *always* have a greater slip angle than at the rear. However, as Figure 3.1 indicates, *excess* understeer must be avoided as it may limit, or even preclude, the vehicle changing direction at high speed.

## Understeer/oversteer, tyre pressures and slip angles

As noted in the previous chapter, *increasing* tyre pressure *decreases* slip angle. *Increasing* loading *increases* slip angle.

Likewise *decreasing* tyre pressure *increases* slip angle. *Decreasing* loading *decreases* slip angle.

If front tyre slip angles always exceed the rear's, that vehicle will safely understeer. If its rear slip angles ever exceed the front's, the vehicle will oversteer. This is controllable by a skilled driver in a solo vehicle, but very hard to correct when towing a heavy caravan. It is far from easy even with a much lighter and shorter camper trailer.

## Understeer/oversteer and mass distribution

Changes in vehicle loading change the extent of understeer/oversteer balance, so suspension engineers build-in a margin of understeer that is maintained up to the vehicle's specified maximum loading. There is, however, a limit to the amount of understeer feasible. An excess may limit or (even preclude) making a desired strong emergency swerve.

If a vehicle is rear-overloaded or has overly stiffened rear suspension, and yaws strongly at speed, it can be suddenly triggered from understeer to oversteer, i.e. its margin of understeer is lost. If that happens at speed the rig may sway out of control and jackknife. (That speed too is a major factor is explained in Chapters Four and Six.)

## Dirt roads

Most of the above relates to hard-surfaced roads. Driving on dirt roads is like skating on ball bearings: tyres at normal road pressures behave much as if they were solid rubber and if cornered too strongly will slide. Experienced outback drivers accordingly reduce tyre pressures by about 25% and keep speed below 80 km/h.

Caution is particularly necessary whilst towing on such roads. Heavily laden tow vehicles will tend to slide at their rear, particularly if excessively end-heavy. A caravan too is likely to slide whilst cornering.

Broome police have clocked caravans being towed at 135 km/h on three-lane dirt stretches of the Cape Leveque Road, despite a big sign stating 'Inadvised for Caravans'.

Many caravanners ignore this - despite that road being notorious for high-speed rollovers - and drivers attempting the 450 km return trip (from Broome) in one ultra-long day.

## Summary

Once how tyres actually behave (re understeer/oversteer) is understood, interactions between caravans and tow vehicles becomes clearer. Or at least less obscure – it is *never* simple!

*Figure 3.2. The Cape Leveque Road (near Broome WA). Pic: AL-KO.*

# CHAPTER FOUR

## Further causes of instability

As emphasised in Chapter Three, the main issue in tow vehicle and caravan handling is an essential margin of maintained understeer.

Speed, relative to the rig's stability, is the major determinant of the upper limit of towing stability. This is particularly so in Australia, where limits are 20 km/h higher than the UK and EU.

Authorities world-wide advise never to tow any trailer towed via an overhung hitch above 60 miles per hour (about 96 km/h). The greater that hitch overhang, relative to the tow vehicle's wheelbase, however, the less stable the rig, and the lower that safe speed.

## Caravan length vs. weight

Tow-vehicle/caravan weight ratio too is a major determinant. It profoundly affects that speed. It is, however not so much a trailer's total *weight* as such that matters but its *length*, and where along that length weight is distributed. That is not just the weight of internal fittings, but very much that of its subsequent loading.

A 4 to 4.5 metre centrally laden caravan weighing 2500 kg is far more stable than a 6 metre caravan of the same weight and manner of loading: its critical speed is higher. It is all but unknown for a sanely laden camper-trailer to yaw or jackknife, despite many exceeding two tonne.

An ideal trailer (stability-wise) has its mass concentrated as close as possible over and either side of a well set-back axle. This reduces its so-called yaw inertia, (inertia is the reluctance of mass to move, or once moving, its reluctance to cease moving). Moreover, the further that mass is located from the tow-hitch, and closer to the axle/s, the better.

UK/EU on-road caravans are typically shorter and 30% to 40% lighter than most local products, and far less end-heavy. Nevertheless, in the UK, and most EU countries, laden caravans are limited to about 80% of the weight of the laden tow vehicle and speed limits are typically 80 km/h.

## Water-tank location

*Figure 4.1. Many recent rollovers involve caravans with forward of the axle/s water tanks. Pic: Clayton's Towing.*

It is vital that trailer mass distribution does not change much with loading. This is a particular issue with changing water tank/s level, unless the tanks are located equally either side of the caravan's axle/s and drawn upon equally. It became a serious issue during 2017-2018. Some long and heavy caravans were made with tanks only in front of the axle/s and relied on those tanks being full to provide tow ball mass. One type (since modified), had only 4% tow ball mass when those tanks were empty.

## Tow-hitch overhang and tow vehicle wheelbase

Tow-hitch overhang is the distance from the tow-vehicle's rear wheel centre line to the articulating part of the tow-hitch. That overhang is the major cause of conventional caravan instability.

*Figure 4.2. As with this hitch (that is well over two metres behind its rear axle), such overhang seriously prejudices stability. Pic: rvbooks.com.au*

Unless one has a fifth-wheeler and suitable tow vehicle, conventional vehicle design makes hitch overhang inevitable, but the shorter that overhang the better. The average (in Australia) is 1.24 metres but some are over two metres.

## Tow-hitch inserts

A tow-hitch insert is held within a tow-hitch receiver. Some inserts, however, have a longer than necessary overhang.

*Figure 4.3. Many hitch inserts have grossly excess overhang. Pic: rvbooks.com.au*

Many owners aware of this issue and its implications have a machine shop shorten the overhang (by drilling a new locating hole).

## Tow ball couplings

Some tow couplings that are longer than average, are claimed to ease coupling and uncoupling, but towing stability is arguably more important. Hitches that are easy to couple and uncouple (yet have minimal overhang) are readily available anyway.

*Figure 4.4. This tow hitch has minimal overhang and is easy to couple and uncouple. Pic: Hitch-EZY.*

Equally important is the hitch overhang's ratio to the distance between the tow-vehicle's front and rear axle (wheelbase).

If feasible, buy a tow vehicle that has the longest wheelbase, the shortest rear overhang and is heavier than most when fully laden.

That of most 4WD tow-vehicles in Australia vary in wheelbase from 2.75 metres to 3.2 metres, hitch overhang averages 1.24 metres (an average ratio of 2.42).

Some (but not all) Land Rovers have about 3 metre wheelbase and under 900 mm hitch overhang (a ratio of 3.33). The greater that ratio, the more stable the rig and the higher its critical speed.

Most US imports such as the Ford 250 and Dodge RAM have a very long wheelbase (up to 4.0 m). It is that long wheelbase, not just their heavier weight, that makes them good for towing.

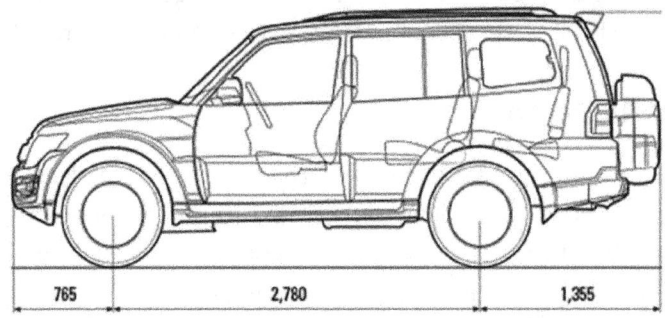

*Figure 4.5. Important for tow vehicle stability is the ratio of the tow vehicle's wheelbase (here 2.780 m, to its hitch overhang of a probable 1.3 m) of a 4WD Pajero. Sketch: Mitsubishi Motors.*

## Tow ball mass

To limit instability, caravans *must* be front heavy. Furthermore, long caravans need proportionally more tow ball mass than short ones. A good general rule (for the generally heavier Australian-made caravans) is that tow ball mass be 10% of the overall laden weight. Those under 4.5 metres, particularly camper trailers, need only 7%. See also Chapter Four.

Tow ball mass and stability are also *speed*-related. The higher the towing speed the more important to have the appropriate tow ball mass.

## Towing capacity

Towing capacity is sadly almost universally misunderstood, because vehicle makers rarely explain what their 'towing capacity' truly means.

Most 4WDs are made for mining and military use - not towing caravans. Many used for towing have a claimed towing capacity (in Australia) of 3500 kg. What that really quantifies is the ability to do so pulling 3500 kg via a length of rope, or a 3500 kg dog trailer.

Vehicle makers quantify towing capacity by their product's ability to stop and restart (in forward and reverse) on a specified gradient (typically 14%), and the ability of its structure to withstand the load etc.

Such quantification is valid for fifth-wheelers (that have the hitch over the tow vehicle's rear axle/s): but does not imply it is feasible to tow a 3500 kg caravan via an overhung hitch.

## Gross Combination Mass

The choice of caravan is commonly restricted by Gross Combination Mass (GCM): the maximum permitted mass of the combined laden tow vehicle and laden trailer (set by the tow vehicle maker).

If unaware of this legal restriction you may well exceed tow-vehicle payload when towing your caravan of choice.

This issue often affects dual-cab utes. With most such, using anything like their full payload results in the rig exceeding the makers' GCM. It may also result in an owner using a minimally laden tow vehicle to tow a now proportionately-heavier laden caravan.

The higher the towing speed the more important it is to have the tow ball mass required. This was prejudiced (particularly in Australia around 2015) when many tow vehicle makers reduced permitted tow ball mass: typically from 350 kg to 250 kg.

Some caravan makers correspondingly reduced their recommended tow-ball mass with no apparent changes to their products - seemingly without regard for the underlying physics.

As noted previously, a few caravans had a totally inadequate 4% tow ball mass unless the front-located water tanks were full whilst towing. For a long and heavy caravan that 4% tow ball mass may reduce the rig's critical speed to well below Australia's towing speed limit. Some such caravans have since been modified. Those that have not, should be, or towed only with full tanks.

## Tow ball mass and weight distribution hitches

Via the lever effect introduced by an overhung hitch, a caravan's essentially-required tow ball mass pushes down on the extreme rear of the tow vehicle. This causes that vehicle's front end to lift, thereby reducing the weight on its front wheels. That, in turn, reduces their slip angles and hence their margin of understeer. That this was undesirable was recognised (in Australia in 1950).

# Weight Distribution Hitch (WDH)

Inventor Mr Haymen Reese developed what, in effect is a springy beam (that still enabled turning) between the caravan and its tow vehicle. His resultant Weight Distribution Hitch (WDH) levers up the rear of the tow vehicle (thereby reducing some of the weight borne by its rear tyres) and restoring some or all of the weight on its front wheels. It also transfers a minor down force to the caravan's tyres.

This effect is often misunderstood. Despite a WDH vendor once claiming otherwise, a WDH cannot move *mass*: it moves the gravity-caused down-force on mass that we refer to as 'weight'.

*Figure 4.6. Typical Weight Distribution Hitch. Pic: Hayman Reese*

# WDH - pluses and minuses

A WDH usefully assists to restore the tow vehicle's weight balance. Whilst it effectively lessens the weight imposed by tow ball mass on the tow-vehicle's rear tyres, it cannot reduce existing yaw forces.

With some vehicles and hitches, adding a WDH *increases* tow hitch overhang, and hence yaw. As a direct result, the tow vehicle' rear tyre slip angles are undesirably increased - yet yaw forces remain unchanged or even increased.

Whilst a WDH is all-but essential when towing a caravan that is substantially heavier than the tow vehicle, that WDH inherently reduces the rig's margin of understeer. As explained below, the extent by which that happens can be partially limited. This situation is not, however, desirable.

*Figure 4.7. Whilst this caravan has (desirable) rear-located axles, it needs a weight distributing hitch to restore weight over the tow vehicle's front tyres: Pic: original source unknown.*

As with the Caravan Council of Australia, RV Books strongly recommends not to tow a caravan longer than 4-4.5 metre, that when laden, is heavier than the laden tow vehicle. No WDH is then required, nor desirable.

Reducing the tow-vehicle's front tyre pressures restores the otherwise minor loss of understeer.

## WDH adjustment

Many people tow caravans that are heavier than whatever tows them. This is inadvisable but, if the tow vehicle maker permits that (not all do), a WDH is then all but essential or too much weight (down force) is removed from its front tyres, thus reducing its margin of understeer.

## Suspension modifications (tow vehicle)

A vehicle's suspension is designed to carry a specific maximum weight. Its margin of understeer is reduced when laden to its maximum, so car makers may suggest a higher rear tyre pressure be used to ensure it will not oversteer. This is essential when towing. Chapter Two suggests the amounts.

*Figure 4.8. Measure the height of the tow vehicle's front wheel arch prior to, and then after, coupling up the caravan (with the WDH not in use). Then adjust the WDH unit's tension such that about 50% of that height difference is restored. Do not to attempt to restore more than that. This will leave the rear of the tow vehicle slightly down. Do not attempt to restore that droop. Pic and instructions: Cequent.*

It is essential to ensure the full range of rear spring travel is not exceeded, such that the axle does not reach its bump stops (below), but that rarely happens unless that vehicle's rear is truly overladen.

Whilst many owners stiffen the rear springs or further inflate air bags, doing so reduces the vehicle's margin of understeer. Doing either causes the rear tyres to increase slip angle - towards possible oversteer.

If your rear suspension has settled to a level lower than when it was new, it's odds-on it is being (or has been) overloaded. A spring repairer, can readily reset those springs. If you **still** wish to stiffen them, doing so equally front *and* rear ensures vehicle handling is not degraded.

## Airbag issues

Leaf springs and basic coil springs depress in proportion to the imposed load. Airbags are not like this: they are very soft at low to medium pressures, but become almost rock-hard once past above 70% or so of their full deflection. Airbags are extremely effective, as their pressure can be reduced when not towing, but must be of the right size for your vehicle and used only at the recommended pressures. They can become rock-hard if over-deflected.

*Figure 4.9. Typical rubber bump stop. It is shaped such it progressively compresses. Pic: Original source unknown.*

If airbags are fitted to the rear of fully-laden utes there is a risk of bending their chassis if the air bags (or any springs) are fully deflected. This particularly occurs when traversing a raised cattle grid at speed. The caravan will pitch strongly fore and aft - imposing a huge down-load on the tow vehicle's rear suspension.

*Figure 4.10. High quality airbag.
Pic: Loadlifter (USA).*

This particularly affects dual-cab utes as the chassis, rear of their cab, has no monocoque structure to react such loads - yet is too short to flex to absorb them. There is thus a stress concentration that is not helped by most makers reducing chassis steel thickness from 3.5 mm to 3.0 mm around 2015.

Many factors affect caravan and tow vehicle stability. All, in one way or another, affect the understeer margin and the rig's critical speed.

Above all, understeer must never be reduced to the extent that it becomes oversteer. This is a particularly serious problem whilst towing as an oversteering vehicle towing a conventional caravan can, within a second or two, escalate to jackknifing. Almost all dash-cam videos of rollovers show that rapid build-up to terminal oversteer.

*Figure 4.11. Dual cab utes used to tow heavy caravans are prone to have bent chassis. Pic: RVdaily.com.*

## Independent suspension

Independent suspension is only necessary for *steered* wheels where a soft ride is desired (it is rare for trucks to have it). It is not necessary for unsteered wheels (Chapter Five explains why. Look underneath most (even top-end) 4WDs and you are likely to find a rear beam axle

It is likewise unnecessary (and undesirable) to have ultra-soft long travel caravan suspension, not least as that mostly used introduces undesirable roll-steer.

## Caravan independent suspension

There are two main reasons why many local caravan makers use independent suspension. The main one is marketing - to the effect that whilst a beam rear axle is just fine for passengers travelling in whatever 4WD tows it, a caravan's suspension must somehow have far more travel and be softer. This is despite any number of imported caravans having the totally adequate (for on-road use) limited travel AL-KO rubber suspension - Figure 4.13.

A well-suspended beam axle is technically superior for trailer suspension, but no major local manufacturer (as far as is known) makes a truly well-engineered beam axle and spring product for caravans.

*Figure 4.12. 2019 Ford Ranger rear beam axle.*
*Pic: Ford Motors*

AL-KO's rubber-in-torsion independent suspension, however, is totally adequate in Australia. It has limited vertical travel but is just fine on-road. It is rugged (and scalable for the caravan's laden weight) yet light and affordable: ideal if building a caravan oneself.

## Yaw (sway) reduction

The early transport industry addressed yaw by removing its cause: of tow hitch overhang. Fifth-wheel caravans, however, need a specialised tow vehicle, unsuited for other use. Because of this, the major market has long been for conventional (pig-trailer-based) caravans, resulting in a substantial manufacturing industry that addresses its inherent failings in various ways and degrees of effectiveness.

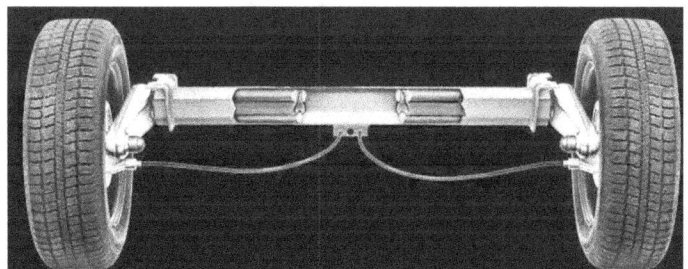

*Figure 4.13. AL-KO rubber-based suspension. Pic: AL-KO*

## Friction yaw reduction

The simplest form of reducing yaw is the AL-KO friction-lined tow ball coupling. It reduces minor yaw that annoys but is not dangerous. Yaw forces, however, increase four times for each doubling in speed. Friction damping assists - but at 100 km/h by only 1% or so. Because of this, most UK/EU caravans now use that hitch to limit 'nuisance yawing', then add electronic sway control (Figures 4.15 and 4.16) for high speed protection.

## Sprung dual-cam yaw reduction

Sprung dual-cam yaw reducers lock the tow vehicle and caravan in a straight line. The rig can only corner by distorting the tow vehicle' tyre contact patches. The cams normally only unlock for tight turns at low speed. These system are very effective in normal driving. If, however, the cams release to enable a high-speed emergency swerve – yaw force energy is suddenly added to the disturbing energy forces - when least needed.

*Figure 4.14. The AL-KO friction tow hitch. Pic: AL-KO (UK)*

## Electronic stability control - Al-KO ESC

This system senses one major yaw cycle or four successive lesser yaw cycles that exceed a lateral yaw acceleration of 0.4 and 0.2 g respectively. If either occurs, a mechanism applies the caravan's brakes at 75% of full braking in three short bursts, straightening the rig and reducing its speed. This cycle keeps repeating if necessary.

*Figure 4.15. Reese dual cam sway system. Pic: Reese*

In essence AL-KO ESC is a 'parachute' approach acting as a rarely-needed but effective back-stop for normally stable caravans. It has the advantage that it does not mask minor yaw (that is best reduced at source anyway).

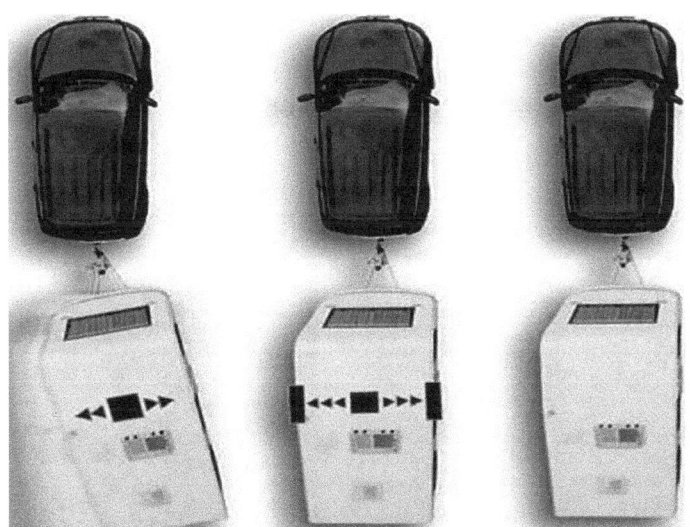

*Figure 4.16. AL-KO electronic stability control. Pic: AL-KO*

## Dexter DSC

This system actuates at lower levels of yaw. As with AL-KO ESC, it actuates the trailer brakes, but in a different manner.

If the trailer yaws to the left, the DSC system actuates its left-side brake/s only. This straightens the trailer more effectively, but as only one wheel (or pair if a twin axle caravan with all four wheels braked), it cannot reduce the rig's speed as quickly as AL-KO's ESC system.

The Dexter system works all the time, reducing the need for friction mechanisms. A possible downside (in RV Books' opinion) is that this may mask inherent instability that may only manifest in a rare emergency swerve. Such issues are best addressed at source. This system is often seen as a good choice for drivers with little or no towing experience, the AL-KO ESC system better suits experienced caravaneers able to reliably control a swaying rig.

Both AL-KO and Dexter responsibly advise that their systems cannot overcome the laws of physics. In this case the major one is the grip of those hand-sized tyre contact patches.

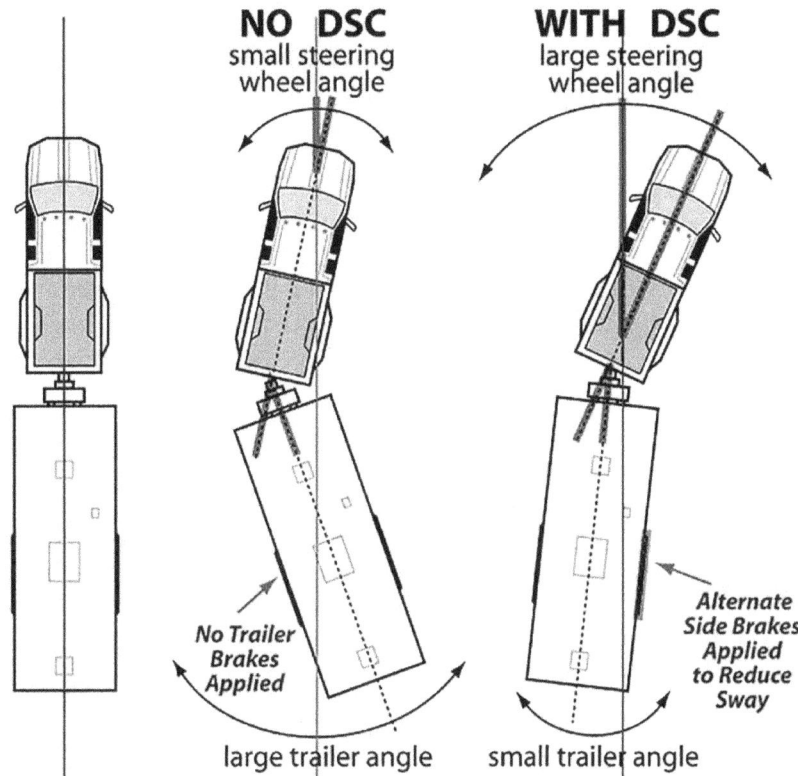

*Figure 4.17. Dexter DSC also corrects minor sway. Pic: Dexter.*

Neither company claims their product will correct a rig once a jackknifing sequence has been triggered if above the rig's critical speed. In a borderline situation, such systems may catch it in time to prevent it by reducing speed below that critical (but neither company claims that).

Most promotional claims and associated video sequences are limited to testing at just under 100 km/h.

## Cruise control risk

When yawing occurs, the associated energy required causes an inevitable minor drop in speed. If cruise control is engaged at the time it will accelerate the vehicle to restore the set speed. This feeds energy into the already unstable system – thus increasing the yaw forces.

Meanwhile, the tow-vehicle's tyres heat up and slip angles increase. Whilst convenient, using cruise control whilst towing a heavy rig at speed is not a clever thing to do.

## Side wind gusts from passing trucks (and when passing trucks)

Trucks generate a strong buffeting side wind from large frontal area, and particularly strong from those with a bluff front (i.e. no long bonnet). There are varying sequences and consequences.

Figures 4.18, 4.19 and 4.20 show the sequence when a truck (particularly if towing trailer/s) is passing a tow vehicle and caravan. As can be seen, that tow vehicle's overhung hitch virtually sets up and then reinforces a yaw sequence that alternately swings the caravan and tow vehicle in opposite planes.

This initiates rapidly-escalating yawing. It is virtually a *recipe* for jackknifing.

These pics are reproduced by kind permission of caravanchronicles.com

They are *strictly copyright* of that website.

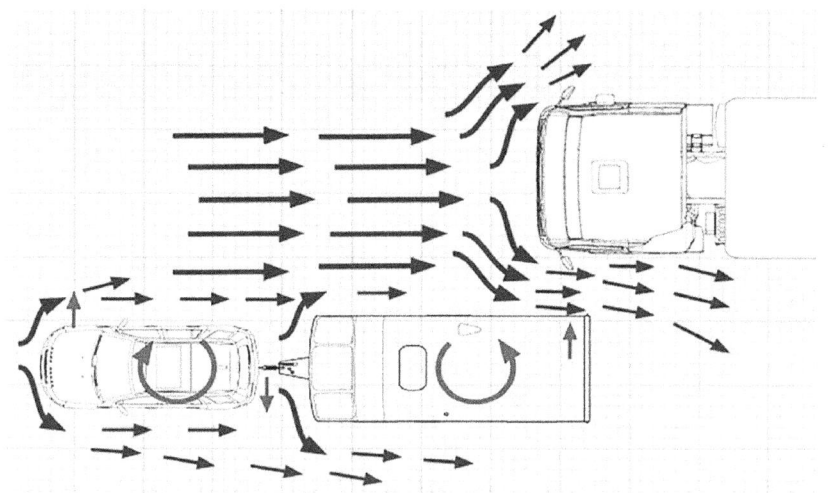

*Figure 4.18. A truck begins to overtake, but is still to the rear of the caravan's axle/s. Vortex and other effects swing the caravan anti-clockwise and (due to hitch overhang) the tow vehicle clockwise.*
*Pic: (copyright)* caravanchronicles.com

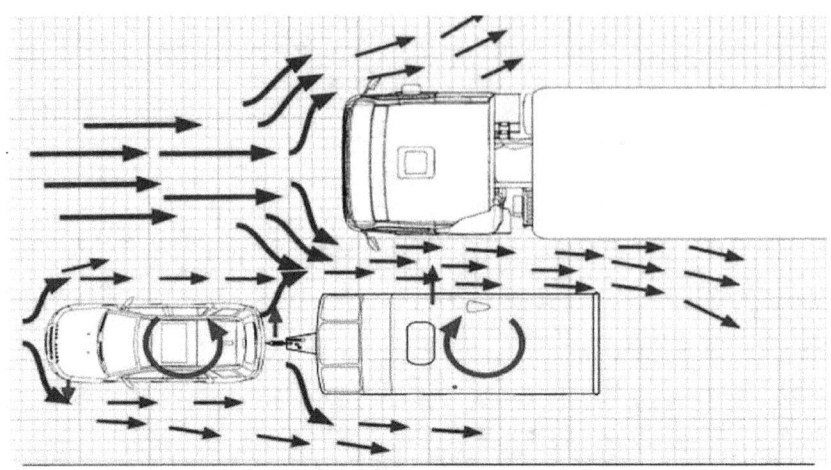

*Figure 4.19. The truck is now ahead of the caravan's axle/s. Vortex and other effects now swing the caravan clockwise. Hitch overhang causes those forces to swing the tow vehicle anti-clockwise. Pic: (copyright) caravanchronicles.com*

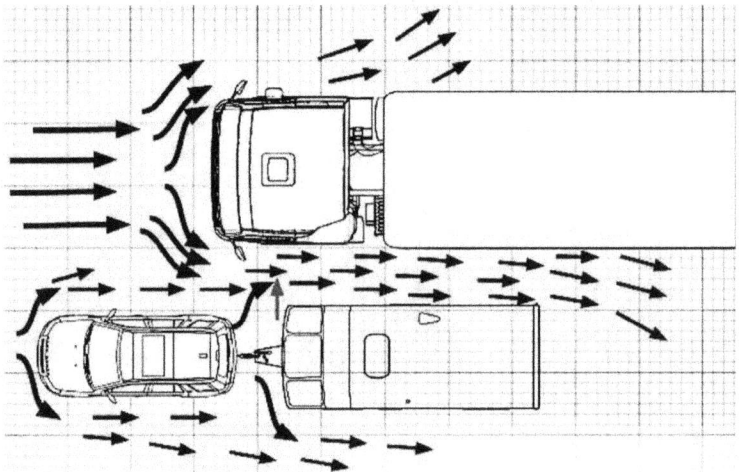

*Figure 4.20. Truck bow is about to overtake the rig's tow vehicle. As the red arrow shows, vortex and other forces are now driving the rear of the tow vehicle toward the overtaking truck, thereby swinging the caravan clockwise, and (again due to hitch overhang) the tow vehicle anticlockwise. This has now set up a sequence that may well end-up jackknifing the rig after few more escalating cycles. Pic: (copyright) caravanchronicles.com*

## Driver skill and reactions

Driver skill here is recognising and understanding this sequence is far from uncommon, and if feasible, slowing down. Travelling at the same speed as heavy transport assists to avoid overtaking and being overtaken, but is only safe with truly stable tow vehicles and caravans.

When overtaking a slow moving vehicle, only do so when there is adequate distance, so that you do not have to increase speed to over 100 km/h. Then leave as wide a (sideways) gap as possible.

That a driver (in Australia) with a basic car driving licence may legally tow many tonnes of potentially unstable trailer at 100 km/h without prior training defies sanity. Both RV Books and the Caravan Council of Australia strongly recommend that first-time caravan buyers' complete an accredited towing course. Such courses cannot substitute for experience, but will warn of the risks.

## Loading your rig

As far as its loading permits, carry anything heavy in the tow vehicle. This is not possible with many dual cab utes when (unadvisedly) towing caravans that are heavier than that ute, but load that vehicle right up to the permissible limit constrained by its Gross Vehicle Mass and Gross Combination Mass. (These terms are explained in Chapter Seven).

Many owners wrongly (and dangerously) believe that as long as one has the required nose mass, where you load things along a caravan's chassis does not matter

Some take this to extremes: one known example (whose owner approached rvbooks.com.au for advice) had a 350 kg set of weights in a front boot partially 'balanced' by a 250 kg motorcycle on a massively heavy rear bumper bar. This resulted in the caravan resisting changing direction when it was needed to, and yawing strongly above 60 km/h.

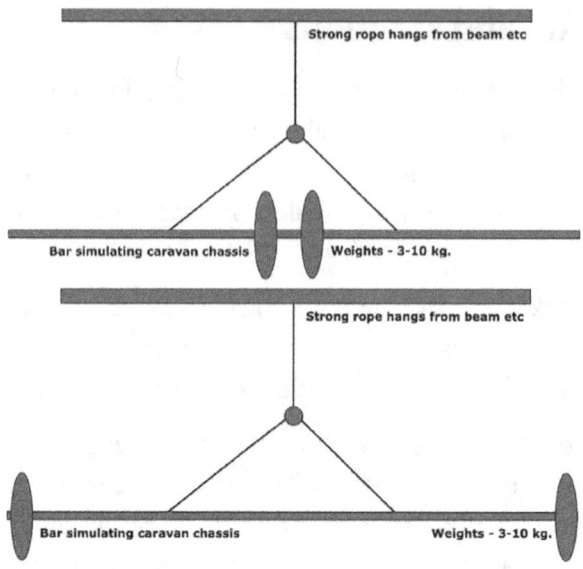

*Figure 4.21. Centralised and heavy end weight. Top: the weights are close to the bar's centre. The bar can be rotated and stopped from rotating with ease. Below: End weight renders that bar hard to start rotating and even harder to stop it. It is related to the hammer thrower effect. Pix: rvbooks.com.au.*

Another, that never travelled off-road, carried some three months supply of canned food (about 150 kg) and over 100 litres of water in 15 litre plastic casks (100 kg plus).

Civilisation extends beyond the end of freeways. Remote towns and Aboriginal communities have supermarkets that stock virtually anything you can buy in town (except edible bread). There is rarely any need to carry more than one week's supplies.

# Part one summary

## Tow vehicle and caravan behaviour

This is well understood, both in theory and practice. It is essentially to ensure the tow vehicle cannot (especially with caravan attached) *ever* be triggered from its essential margin of understeer - to even *momentary* oversteer.

The major factors needed to ensure this cannot happen include:

A long-wheelbase tow vehicle is more preferable than a short-wheelbase tow vehicle, and its hitch overhang be as short as possible. If buying a new tow vehicle both should be a major consideration.

The laden tow vehicle should weigh at least as much as the laden caravan: preferably heavier by 20-30%. Many owners have the opposite, but this is risky unless speed is kept well below towing speed limits. They may get away with it for years, but primarily as they have not needed to make a very strong emergency swerve, or experienced a strong side-wind gust at speed.

Locate heavy personal effects in the tow vehicle, or centrally in the caravan. Never locate anything heavy, such as spare wheels or tool boxes, on the rear of the caravan. If they are there now, remove them, and, if feasible, carry them in the tow vehicle - or locate them on a fold-down cradle under the caravan, close to its rear axle.

*Figure 4.22. Readily-made chassis spare wheel carrier. Pic: Original source unknown.*

Water tanks should be wide (but not long), have internal baffles and be located centrally. If they are front mounted, tow ball mass will vary, (potentially dangerously) depending on the amount of water in them at any time. If yours are like that, consult the Caravan Council of Australia(https://www.caravancouncil.com.au) re having the maker relocate them.

Friction sway devices smooth minor snaking at low and medium speed, but have next to no effect at 100 km/h.

Dual cam devices work well but, if overwhelmed, suddenly release disturbing forces when least needed. If one is required to limit yawing it is better to limit or (preferably) remove the causes of that yaw.

Correct tow vehicle tyre pressures are essential. A major cause of otherwise stable rig swaying is not realising it may be due to grossly under-inflated tow vehicle rear tyres. It also helps to reduce the pressure of tow vehicle's front tyres as less weight is carried by them whilst towing. By and large the tow vehicle's rear tyres should be 50-70 kPa (7-10 psi) higher than the front tyres. Unless this is assured (and particularly if a WDH is used) the essential margin of understeer is compromised - or even lost.

Caravan tyre pressures should be that recommended by the tyre maker for the load carried. This cannot be done without weighing the vehicle on a weighbridge. Do not attempt to guess this. Ongoing police checks show that almost all RV owners substantially underestimate their rigs' laden weight, and some grossly so. Measure the laden caravan and laden tow vehicle on a certified weighbridge, both coupled up and separately.

Tyres with good sidewall stability (such as light truck tyres) assist the stability of both the tow vehicle and caravan. Here too, that air pressure ratio for the tow vehicle's front/rear tyres is essential.

Heavy stuff (such as the kitchen, water tanks, batteries etc.) should be located as close to the axle/s centre line as feasible, and such that this results in the required tow ball mass. When purchasing a new caravan seek one that is designed like that. Avoid those with rear kitchens.

With tow vehicles, winches, bull bars, roof racks, fridges etc all add to their weight. It is only too easy to add an unexpected 200 plus kg.

When doing any/all of this keep an ongoing check of tow ball mass as that still needs to be as close to 10% as possible.

# PART TWO

## Towing stability - how does your rig rate?

As emphasised throughout this book, most reasonably balanced caravans and tow vehicles behave well under normal driving conditions. That vital to towing safety, however, is how your rig will behave in an emergency.

How will it behave if you need to swerve at speed to avoid a head-on collision? Or when being passed closely on a windy freeway by a B-double travelling at speed.

That needed by any rig is an adequately sufficient margin of stability: in effect that your rig remains stable in any reasonably foreseeable situation.

The following is a *very* approximate guide to the probable stability of your fully laden caravan and tow vehicle in emergency situations. It is not feasible practicably or legally to give estimates of safe speeds but the guide assists pinpointing areas that need attention.

A further major factor for a caravan is that point along its chassis where the centre of mass would be, were the entire mass to be concentrated in one place. A caravan must be as centrally loaded as possible.

To use this guide, it is first necessary to know both the tow vehicle and the caravan fully laden weight. Have that checked on a certified weighbridge. Do not attempt to 'guess' this. Police checks show that of all caravan rigs checked, only 4% knew what it *actually* was. Almost all were are overweight, a few grossly so (one by 400 kg).

Those weights known, fill in the data on the seven sections (of the questionnaire) as they relate to your rig.

Then assess the score for each section. Add those scores together - and check the results at the end.

# Caravan Stability Questionnaire

**Ratio of laden caravan weight to laden tow vehicle weight**

*Score:*

| 10 | if the laden caravan weight is 80% or so less than that of the laden tow vehicle. |
|---|---|
| 9 | if that weight is 90%. |
| 8 | if that weight is equal to or under 5% more. |
| 0 | if *more* than the weight of the laden tow vehicle (or I do not know and have guessed). |

**Wheel base (distance between the tow-vehicle's front and rear axle centrelines. Measure (do not use maker's specifications)**

*Score:*

| 10 | if over 3.5 metre |
|---|---|
| 9 | if between 3.3 – 3.4 metre |
| 8 | if between 3.2 – 3.1 metre |
| 7 | if between 3.1 and 3.0 metre |
| 6 | if between 2.9 and 2.8 metre |
| 5 | if between 2.6 and 2.8 metres |
| 3 | if less than 2.6 metre. |

**Length of tow vehicle hitch overhang (distance from its rear axle centreline to towball). Check with a tape measure.**

*Score:*

| 10 | if under 1.1 metre |
|---|---|
| 9 | if between 1.1 and 1.25 metre |
| 7 | if between 1.25 and 1.5 metre |
| 3 | if between 1.5 and 1.7 metre |
| 0 | 3 if over 1.8 metre. |

## Towball mass (typical Australian-designed caravans over 5.5 metre body length)

*Score:*

| 10 | if the towball mass is 10% or over of the trailer's laden weight. |
|---|---|
| 9 | if the towball mass is about 9% of the trailer's laden weight. |
| 5 | if the towball mass is about 8% of the trailer's laden weight. |
| 4 | if the towball mass is about 7% of the trailer's laden weight. |
| 0 | if the towball mass is 6% or under of the trailer's laden weight (or 'I do not know'). |

*Score:*

| 10 | if loads are mainly centrally located |
|---|---|
| 0 | if any load (such as spare wheel/s, tool box, bicycle/s etc are placed at the extreme rear of the trailer - or externally at the rear. |

*Score:*

| 10 | if Light Truck tyres are used |
|---|---|

*Score:*

| 15 | if your tow vehicle rear tyre pressures are set to 50 kPa (7 psi) or more than the front tyre pressures when towing. |
|---|---|
| 0 | if your tow vehicle rear tyre pressures are not increased when towing. |

*Score:*

| 5 | if you never exceed 100 km/h. |
|---|---|
| 10 | you never exceed 90 km/h. |
| 0 | if ever exceed 100 km/h. |

**Total your scores and see 'Scores' for your results**

<u>*Scores*</u>

If you scored between **50** and **55**, your stability is significantly above average.

If you scored between **45** to **50**, your stability is likely to be above average.

If you scored between **41** to **50**, your stability is about average: there is minor cause for concern.

If you scored between **31** to **40**, there is definitely cause for concern – particularly at over 80 km/h.

If you scored below **30**, your rig needs immediate attention.

# PART THREE

## A more technical explanation

# CHAPTER FIVE

## Suspension

Except for a steam tractor in the late 1800s, Lanchester in 1901, Morgan in 1911 plus Lancia and Dubonnet in the 1920s (that had independent front wheel suspension) until the 1930s most vehicles had beam front and rear axles. Trucks retain beam front axles, and many cars and 4WDs likewise.

*Figure 5.1. Mercedes Unimog portal front axle.
Pic: Mercedes Benz.*

Mercedes Unimog's portal beam axles have exceptional ground clearance (using spur gearing) to gain height. Those, together with a very flexible chassis (and three-point body mounting) enable the tyres to retain ground contact and traction over very rough going.

*Figure 5.2. Unimog campervan conversion showing extreme axle articulation.
Pic: caravancampingsales.com.au*

Such suspension was deemed acceptable for passenger vehicles until the early 1930s. Then, and mainly in the USA, owners began to demand a softer ride. Providing this required softer and longer vertical wheel travel – but doing so uncovered a major issue for beam axle connected steered wheels. Known as gyroscopic precession, doing so resulted in steered front wheels swinging from side to side, meanwhile tramping up and down.

## Gyroscopic precession

Gyroscopic precession is the movement of the axis of a spinning body around another axis due to a torque. Swing a spinning wheel in an arc and it twists to a right angle of that movement. In doing so, precession generates considerable force.

Such precession is why a bicycle stays upright once moving, and even more so a unicycle. It does, however, cause problems when one steerable wheel of a sprung beam axle rises over a bump. To do so, the axle must pivot from the other front wheel, thus causing that rising wheel to move in an arc. If that wheel is also spinning fast it is thereby caused to precess. As those wheels are also connected by a steering rod, this causes the other wheel also to swing sideways and precess.

Unless damped, this movement may escalate such that the whole axle and wheels assembly 'tramps' up and down (to its full travel), and to swing to and fro from one full lock to the other at considerable force.

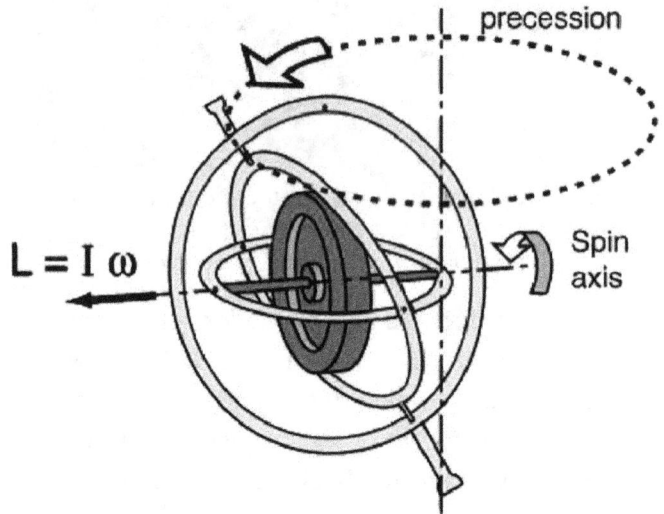

*Figure 5.3. Gyroscopic precession. Pic: enhanced (by adding colour) by rvbooks.com.au from an original - now accredited to many different sources.*

Once commenced that precession may only be stoppable by bringing the vehicle to a virtual standstill. However, reducing speed decreases the tramping frequency but conservation of energy increases the amplitude. This particularly happens if the vehicle has poorly damped and/or very soft suspension long travel suspension.

In the early 1930s General Motors Research Engineer, Maurice Olley noted: *'This torque lifted the down wheel, and slammed its opposite number down on the road in the toed-in position, which continued and built up the cycle'*, (Olley, Chassis Design: Principles and Analysis).

Whilst strong damping assists, it also overly stiffens the ride. Olley realised the solution (for soft long-travel suspension) was such that steered wheels must rise and fall in a straight line: eliminating a common axle can enable this.

The very first known was in 1873. Amedee Bollee's steam driven road vehicle L'Obeissante, had each front wheel suspended via a vertical steel 'cage'. The cage had an elliptical spring on each side to provide suspension movement. Morgan (1909 and Lancia (1923) had steered wheels sliding on what were virtually extended king pins.

Independent front suspension enables a usefully higher sprung/unsprung mass ratio, and a correspondingly softer ride, but necessitates a far more rigid monocoque vehicle body as twisting compromises steering.

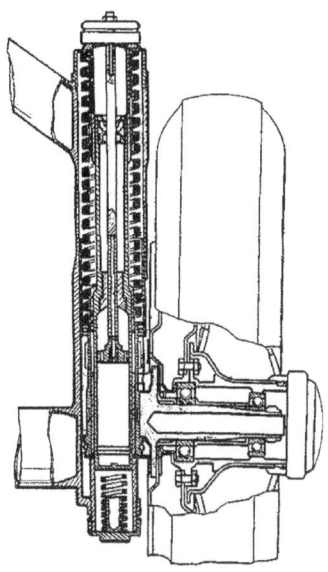

*Figure 5.4. Early (late 1920s) Lancia 'Chantelle' independent front suspension. Pic: Wikipedia commons.*

Passenger vehicle suspension design also has human physiological constraints. It necessitates spring rates of about 1.5 Hz and g force typically limited to < 0.2 g.

## Roll centre height – caravan beam axle suspension

All caravans roll about an axis that, at their tow ball, can *only* be the tow ball receiver. Most 4WD and many cars used for towing have beam rear axles. Thus, no matter how sprung, their roll axis at the rear will, depending on tyre diameter, be about the same height above the road surface as the towball (i.e. 380 to 450 mm).

If as above, this ensures that when the caravan rolls it does so on a virtually horizontal roll axis (at roughly tow ball height) along its full length. This ensures a reasonably stable platform with a roll axis not that much below the laden caravan's centre of gravity. A beam-axled caravan thus rolls only slightly when subject to a side force.

Beam axle suspension also usefully ensures the tyres remain at a right angle to the road. This, in turn, ensures the contact patch is of the horizontal part of the tyre as the tyre designer intended.

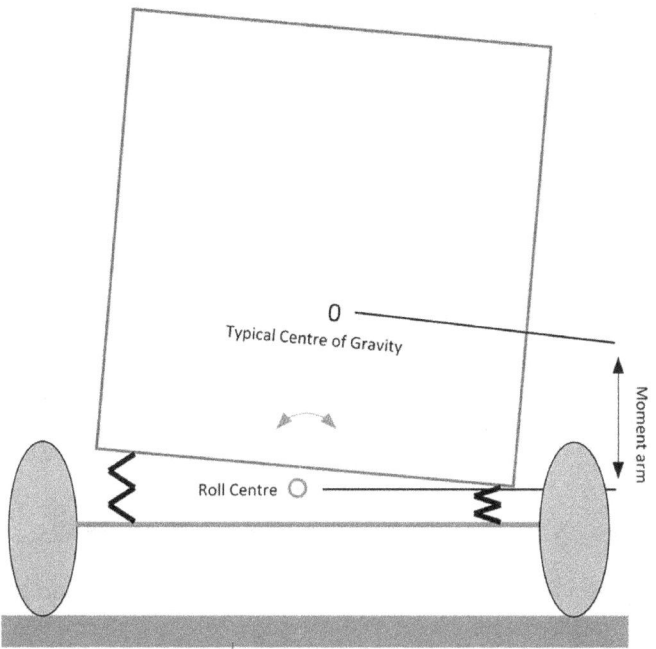

*Figure 5.5a: Beam axle suspension geometry ensure a stable upright platform such the wheels (and particularly tyres) remain vertical as the body rolls. The moment arm (from the cg to the roll centre) is also minimised.*

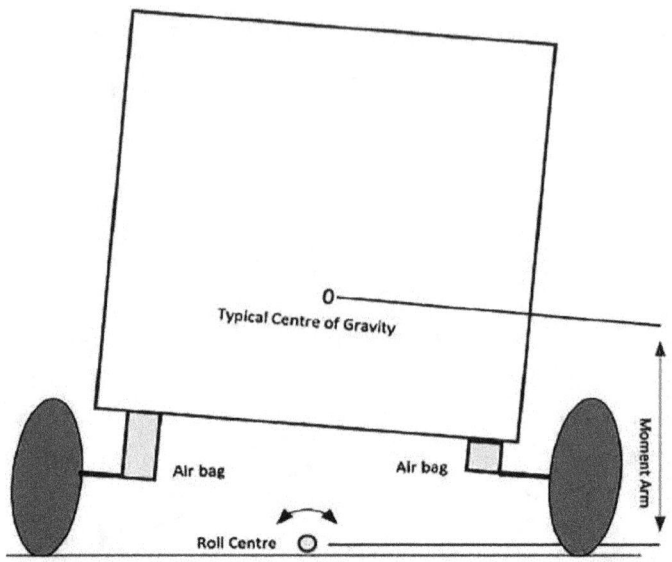

*Figure 5.5b: Trailing arm independent suspension, used for most caravans, enforces a roll axis close to or at ground level. This undesirably extends the moment arm. Further, its geometry causes the wheels to tilt, resulting in the intended tread-only contact patch being compromised. Pix: rvbooks.com.au*

## Roll centre height – independent suspension

Almost all but the twin-beam independent suspension (designed by the late Alan Mawson) results in a roll axis close to or at ground level. Its tow hitch nevertheless can *only* be at tow ball level (380-460 mm high). Any roll thus enforces a sideways and downward diagonal tilt at the axles.

As a direct result of this diagonal roll axis, a caravan with trailing arm suspension, may have its roll centre *below* ground level at its far end. As the centre of gravity remains at the original height this long and undesirable moment arm in the worst possible place, its extreme rear. Despite this, the maker may locate one or two spare wheels half way up it. Plus a toolbox or bicycle rack lower down.

## How much does roll centre height matter?

Caravan and tow vehicle stability has many interdependent variables. If all else is reasonably fine, roll centre height is less of an issue – but many a rig has only border-line stability.

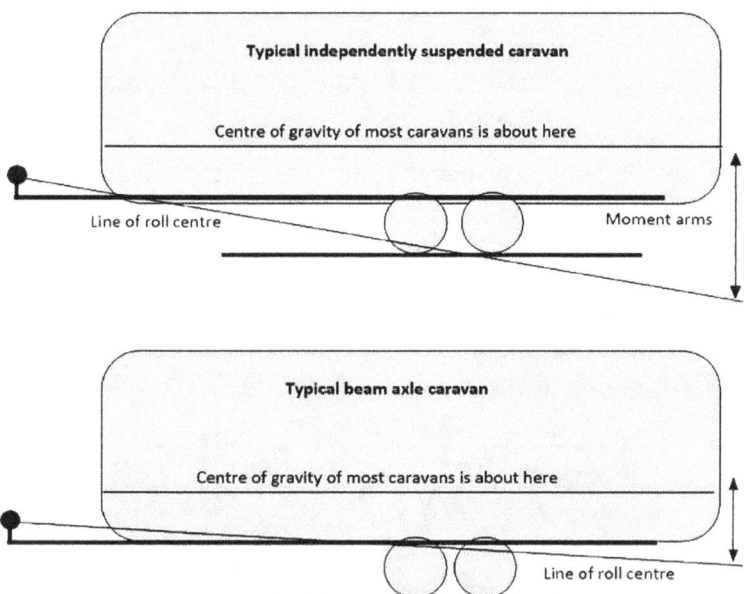

*Figure 5.6. Typical caravan trailing arm independent suspension imposes a ground level roll centre at axle location. The moment arm (relative to its cg) is substantial at its extreme rear. A beam-axle caravan has a desirably higher axis, hence shorter moment arm (relative to its cg. Pix: rvbooks.com.au*

It makes little sense to prejudice stability (and especially critical speed) by totally unneeded long travel trailing arm soft suspension.

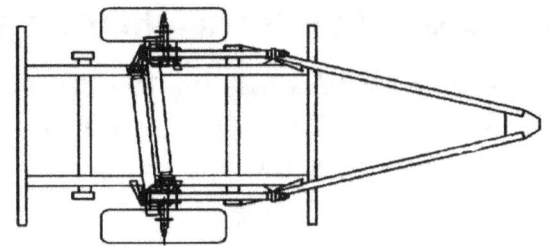

*Figure 5.7. Track Trailer twin beam independent suspension. Pic: Track Trailer.*

A twin pivoting beam suspension was designed by the late Alan Mawson initially for the RAAF's mobile ground radar systems. It is still used by Track Trailer. It has a usefully high roll axis.

A (later) modified version of Alan Mawson's system is used in the Vista Crossover. Both combine independent suspension whilst retaining a desirably high roll centre.

*Figure 5.8. Vista Crossover twin beam independent suspension. Pic: Vista Crossover.*

# CHAPTER SIX

## Critical speed

## The single (fifth-wheeler) pendulum

A fifth-wheel caravan sways like a single horizontal pendulum suspended from the tow vehicle's over-axle located hitch. Its bob, in effect, is the mass of the trailer.

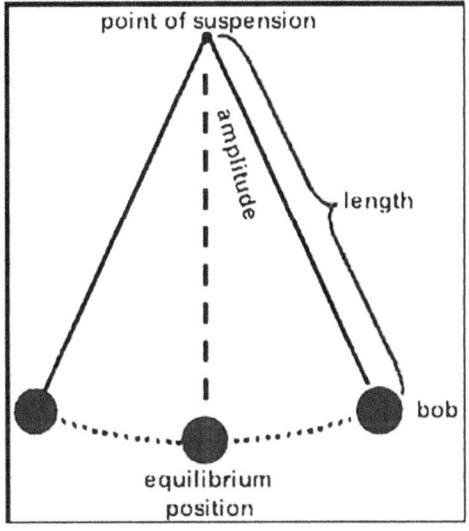

*Figure 6.1. The simple and totally predictable action of single pendulum and thus a fifth wheel caravan and its pivot - the hitch of its tow vehicle.*

Side wind gusts etc may cause that trailer to yaw slightly but the movement is mostly small, quickly self-damped and exerts no forces on the pivot (i.e the tow hitch of the vehicle that tows it). Likewise, if the tow vehicle yaws, it has no adverse effects on the trailer.

As long as the fifth-wheeler's weight on the tow vehicle is within that vehicle's limits, and the hitch is above the tow-vehicle's axle it is stable at any speed.

## Conventional caravan pendulums

A conventional caravan's dynamic action is totally different. In effect it is a double pendulum. The upper pendulum is the tow vehicle and its bob is the overhung tow ball. From that bob hangs a second pendulum (the mass of the caravan).

At low levels of yaw, interaction between the two pendulums is predictable. If the tow vehicle (upper pendulum) yaws to the right, the overhung tow-hitch causes the lower pendulum to yaw to the left. This direction-changing interaction (a 1800 phase-change) is a further cause of conventional caravan instability. The greater that hitch overhang, the greater the (undesirable) effects.

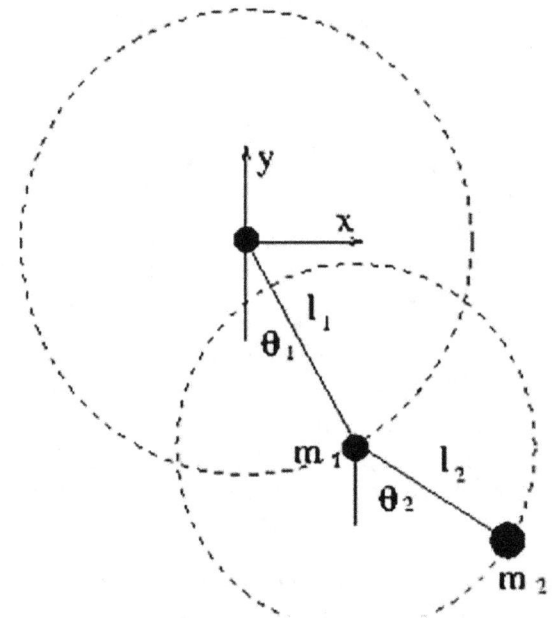

Figure 6.2. *The dynamic behaviour of double pendulums involves double differential Lagrangian Hamiltonian equations. For those interested, see* https://www.myphysicslab.com/pendulum/double-pendulum-en.html.

Double pendulum interaction, however, is hugely complex. It can only be forecast for a very few degrees of swing - determined by the respective pendulums' respective length and mass. Once beyond a specific limit, unique to each pendulum and extent of swing, its interaction suddenly becomes 'chaotic' (it is random-like, but not random).

As experimenting with the example above shows, no matter how careful you are to duplicate the experiment (beyond that minor movement), a double pendulum never follow the same sequence, and the sequences never converge. Such experimenting also shows the seriously adverse effect of extending hitch overhang.

Experimenting also illustrates why a driver cannot correct a strongly swaying caravan above its critical speed. Doing so requires knowing what the rig is likely to do next and making appropriate steering correction, but not possible because ongoing action is random-like.

That also shown is that the 'chaotic' behaviour occurs (suddenly) but only if above a minor degree of movement - below that, swing interaction is predictable and controllable.

In a caravan example, such swing (yaw) is best limited by understanding (or at least accepting) its causes, designing accordingly including adequate damping, such that yaw cannot escalate.

Yaw is primarily damped by the tow vehicle's tyre hysteresis (friction) and aided by tow ball friction etc (as with the AL-KO friction hitch - Figure 4.14). It is mainly of concern if it does not die without additional frictional (or other) aids within two/three cycles.

If yaw continues beyond two/three cycles, it is of serious concern. This is because if that occurs above a critical speed (specific to each rig and its loading), the yaw may self-trigger into a positive feed-back loop fuelled by the energy inherent in the rig's momentum.

UK and EU caravans yaw slightly unless damped, but only minor damping is needed (typically via an AL-KO friction hitch). Most such caravans also have one or another form of electronic yaw control to cope with emergencies. That, (plus low yaw inertia) enables them to need only 5%-6% tow ball mass. None needs a WDH – few (if any) even have provision to fit one.

## Critical speed is rig and loading specific

It has been totally proven both in theory and decades of (sometimes dangerous) controlled trials that there is a critical speed for any specific combination of tow vehicle and caravan and its loading.

That critical speed, and magnitude of yaw acceleration is *directly associated* with the rig's margin of understeer. That in turn is a function of the tow-vehicle's mass relative to the caravan's mass (and particularly mass distribution), length, hitch overhang, tyre type and size, sidewall stiffness and pressure etc.

## Major stability factor

All of the above (and more) is involved but the longer and more end-heavy the caravan (and the lighter the tow vehicle), the lower the speed at which criticality occurs. In essence, the aim is to have the critical speed well above any speed at which the rig is likely to reach.

An excellent demonstration by Professor Jos Darling (of snaking by excess rear end mass above that critical speed) can be seen at https://www.youtube.com/watch?v=zwlgZG55QWk. (It is behind a paywall but the site has a free trial period).

# CHAPTER SEVEN

## Designing a stable pig-trailer caravan

A truly major factor is a conventional caravan's yaw inertia: (i.e. its inertial resistance to yaw and reluctance to cease yawing once that commences). This is associated with its 'radius of gyration' - that point along its chassis where its centre of mass would be were the entire mass to be concentrated in one place.

This can be calculated (theoretically) by cutting a caravan into thin slices that each has a mathematically describable shape and mass. It is feasible to do so - but is a complex process.

A simple, practical and reasonably accurate approach is to place the caravan on a friction-free turntable (Figure 7.1) that is rotated, by about 30 degrees, against the force of two or more springs. It is then released. The time taken to return to centre is then used to calculate the moment of inertia - i.e. 'a length that represents the distance in a rotating system between the point about which it is rotating and the point to or from which a transfer of energy has the maximum effect'.

As far as is known, however, no such table is owned or used by the local caravan industry. If so, that implies that local caravan manufacturers do not know the yaw inertia of their own products.

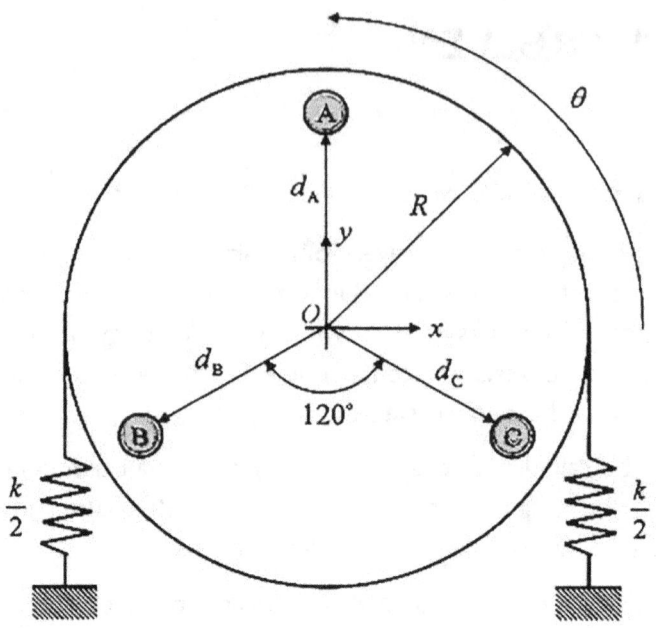

*Figure 7.1. Yaw turntable - simple and effective.
Pic: research.va.gov.*

It has also been suggested that this can be done by literally twisting the caravan on a hard grippy surface and releasing it - using an accelerometer to indirectly measure the yaw inertia.

If the magnitude of the caravan's yaw inertia, plus its 'radius of gyration' (distance from an axis at which the mass of a body may be assumed to be concentrated and at which the moment of inertia will be equal to the moment of inertia of the actual mass about the axis), is known, the caravan's resistance to yaw, and to cease yawing, can be calculated. This is currently not feasible because (as far as is known) no local caravan makers have the facilities to measure this.

Reducing yaw inertia requires the axle/s to be as far back as possible. This is feasible by locating all heavy stuff (such as the kitchen, water tanks, batteries etc) close to the axle/s centreline: and such that this results in 10% tow ball mass (now recommended also in SAE J2807.)

The spare wheel is best located on a swing-down cradle underneath the caravan and as close the axle/s as feasible - as shown in Chapter Five.

LiFePO4 batteries save space and weight and volume: both are about one third that of lead-acid batteries of the same nominal capacity, and one quarter of that in terms of ongoing usable capacity.

RV Books suggests it best not to include spacious luggage area at the caravan's front (and never at the rear) or storing heavy stuff there prejudices the optimal radius of gyration and yaw inertia. It may also encourage owners to overload their rigs.

## Suspension

The suspension should be such that it has a high roll centre. Doing so ensures the caravan rolls about an axis that is by and large horizontal along the length of the caravan.

For mostly road-going caravans, there is no need for long travel suspension: it prejudices stability; moreover a pig trailer based caravan is free to roll about its tow hitch. That is primarily what it does when (say) traversing the often steep entry and exit angle of most fuel stations. Such suspension need only cater for minor road bumps and dips. Off-road caravans need more, but far less than that currently used.

The AL-KO rubber in torsional compression has limited travel yet has proven fine even over dirt roads. It is, however, advisable to use the otherwise optional shock absorbers.

Another totally adequate suspension is the typically two or three-leaf parabolic leaf springs and an RHS (Rolled Hollow Section) beam axle. Such springs are far longer than the 750 mm or so typical trailer springs. Adequate shock absorbers are essential.

## Tow ball mass

That caravans need 10% tow ball mass is a (safe) generalisation. In practice the desirable tow ball mass is also a function of yaw inertia and the caravan's overall length. A short (say) 4.5 metre centre-heavy caravan can cope with less. Seven percent is usually adequate for those.

*Figure 7.2. Parabolic leaf spring - long and supple. Pic: original source unknown.*

## Stability

Unless it has very high tow ball mass, a caravan towed via an overhung hitch will usually sway slightly at low speed. Unless that sway dies of its own accord in two (or most three) cycles, it needs redesigning until it does. Then and only then should any form of sway control be added, and that is best confined to the AL-KO friction hitch.

Once stability is assured, it is well worth fitting an electronic stability system - but for 'parachute use' only. The Dexter unit operates at lower levels but must *never* be used to 'rectify' a basically unstable caravan.

Caravan dealers should emphasise that their laden products should only be towed by vehicles that are at least as heavy and preferably 20%-30% heavier.

Instruction manuals should warn against (a) overloading, and (b) never exceeding 100 km/h whilst towing.

## Acceleration:

relates to a change in a mass's rate of velocity. It may be positive, or negative as when braking. It is measured by dividing velocity (metres per second) by seconds. That is the same as dividing distance by time squared. The unit often shown as 'G' (but correctly as 'g') is an acceleration of approximately **9.81•m/s²**.

## Ackermann steering geometry:

is a geometric arrangement of linkages in the steering of a car or other vehicle designed to solve the problem of wheels on the inside and outside of a turn needing to trace out circles of different radii.

## Caravanner:

originally spelled caravaner, in 1909 the (UK) Caravan Club changed its spelling to caravanner.

## Energy:

the stored-up ability to perform work (or the capacity for doing work). It can be expressed as force times displacement in unit time. That involved in connection with caravans and tow vehicles is mainly potential energy and kinetic energy - see below.

## Energy (potential):

is the capacity of a body to do work by virtue of its position or configuration. That held within a vehicle spring when compressed is potential energy (also aptly called 'elastic energy'). It may also be the potential energy of water in a dam - that can drive a turbine when released.

### Energy (kinetic):

is that associated with motion. Any moving object is able to do work as a result of it moving. It is often regarded as being the same as momentum (see below), but it is not.

Kinetic energy is proportional to the square of the mass's velocity. It is defined as **KE = ½mv²**. A tow vehicle and caravan travelling at 100 km/h thus has four times the kinetic energy of that same mass travelling at 50 km/h – not twice. The reason why jackknifing is so violent is that the kinetic energy of that rig (as the result of a force that accelerated it to 100 km/h over 100 or more seconds) is released over a final second or two.

### Force:

is that influence on a body that causes it to accelerate. The greater the force applied, the greater the rate of change of acceleration. That rate of change is directly proportional to the force acting upon it. It is inversely proportional to the mass of that body.

Force has magnitude and direction: describing it thus requires both terms.

In SI units, force is expressed in Newtons (N). A Newton is the amount of force required to accelerate 1.0 kg by 1 m/s². That is: **1 N = 1 kg•m/s²**.

### Gross Combination Mass (GCM):

is a legally binding rating set by the vehicle maker of the combined laden weight of that vehicle and the laden weight of whatever it tows. The GCM can be a major determinant of the loading of dual-cab utes when towing a heavy caravan.

**Gross Vehicle Mass (GVM):**

this too is a legally binding rating set by the vehicle maker. It is the maximum allowable laden weight of a motorised vehicle when standing on a public road.

**Inertia:**

is a major factor in caravan and tow vehicle dynamics behaviour. That mass resists changes to its state of motion or rest is known as inertia. It is usually expressed in grams.

**Inertia (moment of):**

is a measure of an object's resistance to changes in rotation (also the capacity of a cross-section to resist bending). See also 'Radius of gyration'.

**Mass:**

this relates to the amount of matter within a body. It is expressed in terms of the resistance of a body to being accelerated.

**Mass and weight:**

these are different concepts. Confusing one with the other may not matter with static objects on Earth. It does, however, for objects that move, both on Earth and in space.

**Moment arm:**

A moment arm is, in effect, leverage. For example, it determines the effect of where weight is located along the length of a caravan relative to location of its axle/s. It also relates to the effect of tow-hitch overhang. See also 'Torque'.

**Momentum:**

the quantity of motion. It is directly proportional to the product of the mass and velocity of a moving object (i.e. momentum = mv).

**Motion – laws of:**

much of caravan and tow vehicle behaviour can be explained by the basic laws of motion defined in the late 1600s by Sir Isaac Newton.

**Motion – first law of:**

unless influenced otherwise by an external force, mass remains at rest or continues to move at constant speed in a straight line unless acted upon by some external force.

**Motion - second law of:**

the time rate of change of momentum (see below) is proportional to the applied force and takes place in the direction that the force is acting.

**Motion - third law of:**

for every action, there is an equal and opposite reaction.

**Power:**

is the amount of work done in a unit of time. For example, the work done when a tow vehicle pulls a caravan up a hill always remains the same. Doing that at 100 km/h, however, will need a lot more power (but same work for a shorter time) than doing so at 50 km/h. A tow vehicle and caravan travelling at 100 km/h has four times the kinetic energy of that same mass at 50 km/h – not twice.

**Radius of gyration:**

this (in mechanics,) is where the centre of mass would be, were the entire mass to be concentrated in one place. More specifically it is the radius of gyration of a body about an axis of rotation expressed as the radial distance of a point from the axis of rotation at which, if the whole mass of the body is assumed to be concentrated, its moment of inertia about the given axis would be the same as with its actual distribution of mass. Mathematicians, however, define it as the root mean square distance of the object's parts from either its centre of mass or a given axis, depending on the relevant application.

The radius of gyration can be calculated by 'cutting a caravan into thin slices' that each has a mathematically describable shape. It can be done more practicably on a friction-free turntable that is rotated by about 30 degrees against the force of two or more springs. It is then released. The time taken to return to centre is then used to calculate the moment of inertia.

**Snaking:**

colloquial term for the lateral high speed oscillation of a towed vehicle: more correctly known as 'yawing'. (See also 'yaw').

**Stability:**

a system that has some form of restoring force. A stable system is one that, if perturbed in some manner from a state of equilibrium, will revert unaided to that stable state. (Understeer is a good and relevant example.)

**Tare Weight (caravan):**

the legally required 'declared' unladen weight of a caravan, 'ready for service' as it leaves the maker's factory. It may not include optional extras (even if included in the original order). These are often supplied and installed by the dealer.

### Torque:

is the effect of an applied force causing something to roll or rotate. It relates to the force itself, and to the distance that force is applied from the axis of rotation.

### Torque and moment arm:

In physics, the terms usually refer to the same thing, but in mechanical engineering torque is the tendency of a force to rotate an object in a pivot: it also a measure of the turning force of an object. 'Moment' refers to the tendency of a force to move an object. A 'moment arm' is the distance between the point of rotation and the force's line of action.

Torque and moment use the (SI) unit 'Newton metre', but torque is shown as Nm/revolution, moment as Nm.

### Tow Ball Mass:

is the essentially required mass imposed on the rear of the towing vehicle via its hitch. The actual amount (typically and desirably 10% of the laden conventional caravan weight) may or may not be recommended by the caravan manufacturer. Its maximum permissible is the lesser of that set by the tow vehicle maker, and the tow hitch maker. Exceeding maximum tow ball mass is illegal.

### Weight (for technical purposes):

mass attracts mass. 'Gravity' is the attractive downward force of the Earth's mass on another mass and expressed as weight. Because no gravitational forces act upon it, mass has zero weight in space. But were, in space, a mass of (say) 5 kg to be thrown at someone, its impact will be much the same as on Earth.

**Work:**

has a technically specific meaning: it is the transfer of energy or the application of force over a distance. Lifting a heavy object and putting it on a high shelf is one example. The unit of work is the Joule. One Joule is one Newton times one metre.

**Yaw:**

refers to a deflection in a rotational or oscillatory movement of a body such as a caravan about its vertical axis. Many people refer to yaw as sway but this can confuse because when a caravan sways (rolls) its centre of gravity moves laterally. When it yaws it generally does not.

**Yaw Force:**

is the influence of (e.g) a side force that causes the front of a caravan to be accelerated sideways. The greater that force, the greater the rate of change of acceleration. As a moment arm (leverage) is introduced by the tow-hitch being located to the rear of the tow-vehicle's rear axle/s that force is increased by the length of the moment arm of that hitch.

**Yaw Inertia:**

can be seen (for the purposes of this book) as the resistance of mass to that change of motion when that mass is subject to a side force ('yaw').

# REFERENCES

## SAE International J8027

**'Performance Requirements for Determining Tow Vehicle Gross Combination Weight Rating and Trailer Weight Rating'**

In 2008, and led by Toyota, the (US) Society of Automotive Engineers released a major set of recommendations (SAE J2807 Recommended Practices) to help vehicles confidently tow a rated maximum trailer weight under a wide range of real-world driving conditions.

It took many years for vehicle manufactures to officially follow these recommendations, but (again) led initially by Toyota, all US motor manufacturers and the three major Japanese makers of vehicles used for towing trailers now follow SAE J2807.

This subsequently caused some disruption in Australia, particularly in 2015, when some of the most popular USA and Japanese vehicles used for towing had their tow ball mass capacity accordingly downgraded.

The J2807 recommendations, that became in effect a Standard, include:

**Davis Dam test**: Tow vehicles must be able to climb the Davis Dam grade (or equivalent simulation), which is a 915 metre elevation change over an 18.3 km stretch of Arizona State Route 68, southeast of Los Vegas. The tow vehicle must not drop below 64.4 km/h with the air conditioning on maximum cooling.

**Acceleration:** Single rear wheel tow vehicles (with the SAE-specified weight) trailer must be able to accelerate on level ground to 30 mph (48.3 km/h in 12 seconds or less, to 60 mph (96.6 km/h) in

30 seconds or less, and from 40 mph (64.4 km/h to 60 mph (96.6 km/h) over level ground in no more than 18 seconds. Dual rear wheel tow vehicles are allowed minor extra time to meet these requirements.

**Launching:** On a 12% grade, the tow vehicle must be able to climb 16 ft (5.0 metres) from a standstill, five times within five minutes in both forward and reverse.

SAE J2807 uses a specific set of assumptions to calculate maximum trailer weight ratings: Vehicles with a GVWR of less than 8,500 lbs (3855 kg) factor in a 150 lb (68 kg) driver and 150 lb (68 kg) passenger in their tow ratings. Vehicles over 8,500 lbs (3855 kg) GVWR add an extra 100 lbs (45.3 kg) for cargo. When considering a J2807-based tow rating for your vehicle, to subtract additional weight for heavier or additional passengers or cargo load from the maximum tow rating.

**Launch and acceleration:** performance on a level road and a 12% upgrade.

**Combined handling performance:** understeer and trailer sway.

**Combined braking performance:** stopping distance and parking brake-hold on grade.

**Structural performance:** for the vehicle, hitch and hitch receiver.

**Maximum trailer weight ratings:** in addition to performance standards, SAE J2807 includes a specific set of assumptions to calculate maximum trailer weight ratings.

For 'light-duty' full-size pickups the Gross Combination Weight rating (the equivalent of the Australian Gross Combination Mass) of vehicles under 8,500 lbs (8340 kg), SAE J2807 assumes that the tow vehicle includes any options that have more than 33% penetration, that there is both a driver and passenger in the vehicle, each weighing (what seems an optimistic) 150 lbs (68 kg), and that tow vehicles include up to 70 lbs (32 kg) of aftermarket hitch equipment (where applicable). Finally, tow ball mass must be 10% of the laden mass.

## Trailer and tow vehicle stability

An important recommendations in SAE J2807 relates to trailer and tow vehicle stability. It accepts that the major factors are understeer/oversteer and adequate reduction or damping of yaw. It addresses this by requiring that (when tested) tow vehicle/trailer rigs are laden for the highest weight permitted and that payload is evenly distributed across the entire trailer such that there is 10% tow ball mass plus/minus 0.5% or plus/minus 5 kg (whichever is the greater).

If a WDH is in use it shall consist of one or two spring bars with a chain at the trailer end. No sway control or other mechanisms shall be used to increase friction or articulation stiffness.

Under these conditions the rig must be able to corner at a lateral acceleration of not less than 0.3 g with 50% of Front Axle Load Restoration (FLAR) with zero decrease in understeer, and likewise at 0.4 g with 50% FLAR.

In requiring this, SAE J2807 indirectly highlights that adding a WDH reduces ultimate cornering power and, to some extent, quantifies it.

The original (and still valid) original draft of SAE J2807 was at one time available on the Internet - but has since been removed.

SAE J2807 can be purchased from: https://sae.org/publications/

# Books

There are innumerable papers on various aspects of this subject – but very few books.

An excellent book on vehicle handling and stability (but alas not trailers) is Chassis Design: Principles and Analysis. The book was compiled from GM Research engineer Maurice Olley's notes (by Milliken and Milliken) 27 years after his death. It is totally valid to this day. ISBN: 0768008263.

Despite its title, Race Car Vehicle Dynamics is a very practical guide to suspension and tyre behaviour. Included is a CD containing relevant vehicle dynamics software. It was written by Douglas L. Milliken, Edward M. Kasprzak, L. Daniel Metz and William F. Milliken. It is published by SAE International with a Product Code of R-280, ISBN: 9780768011272.

A down-to-earth book covering suspension in detail is Bastow D.'s Car Suspension and Handling. It has many illustrations and worked examples. It is published by the Society of Automotive Engineers, Warrendale PA. Published by SAE International with a Product Code of R-318, ISBN: 9780768008722.

The best book on tyre behaviour has to be Professor Hans Pacejka's monumental Tire and Vehicle Dynamics. It covers everything you need to know about pneumatic tires and their impact on vehicle performance, including mathematics modelling and its practical application. The book is, however, substantially mathematical. The last (3rd) edition is 2002. ISBN: 9780080970165.

Another useful reference book is J. Y Wong's Theory of Ground Vehicles, 4th ed. John Wiley and Sons. ISBN 9780470170380.

# Papers

There are many excellent papers in the tow vehicle and trailer dynamics area. Those listed are primarily those most cited, and the author has found valuable over the years. Many of the 'classic' papers were written in the 1970s but are totally valid today.

There are far fewer papers about tyres: possibly due to the tyre industry being ultra-competitive and not willing to share findings.

## Tow Vehicle and Trailer Dynamics

**Anderson R. J,** and **Kurtz E. F,** *Handling Characteristics Simulations of Car –Trailer Systems.* SAE 00545 – 1980.

**BS AU 247: 1993.** *Method of Test For Lateral Stability of Passenger Car/Trailer Combinations.* (British Standards Institution, London, UK).

**Darling, J,** and **Standen, P. M.** *A Study of Caravan Unsteady Aerodynamics.* Proc. Inst Mech. Engrs, Part D. J. Automobile Engineering.

**Deng W,** and **X. Kang.** *Parametric Study on Vehicle Trailer Dynamics for Stability Control.* SAE 2003-01-1321.

**Fratila D,** and **Darling J.** *Simulation of Coupled Car and Caravan Handling Behaviour. Vehicle Systems Dynamics Through Computer Simulation,* ASME Paper WA/MET-9.

**Killer C. J,** *The Dynamics of Towed Vehicles.* This is (an M.Eng thesis) is a good introduction to the topic. It was originally available at http://towingstabilitystudies.co.uk/stability_studies/ but was recently (inexplicably) removed.

**Klein, R. H,** and **Szostak, H. T.** *Effects of Weight Distributing Hitch Torque on Car-Trailer Directional Control and Braking.* DOT HS-803 248, October 1977.

**Klein, R.H. and Szostak, H.T.** *Development of Maximum Allowable Hitch Load Boundaries for Trailer Towing,"* SAE Technical Paper 800157, 1980 J2807

**Sharp R. S,** and **Fernandez, M.A.** *Car-caravan snaking.* Proc. Inst. Mech. Engrs 2002.

## Tyre Dynamics

**Smith N. D,** *Understanding Parameters Influencing Tire Modeling.* Dept of Mechanical Engineering, Colorado State University.

## Websites

Our own RV site: rvbooks.com.au

This contains regularly updated articles on tow vehicle and caravan dynamics. If considering solar for your RV or home see also our companion site: solarbooks.com.au

See also Simon Barlow's accurate and elegant articles in this area: they primarily relate to UK/EU caravans but are well worth reading: caravanchronicles.com/tag/dynamics-of-towing.

# ACKNOWLEDGEMENT

My writings and this book summarise current thinking. They stem from my interest and involvement whilst employed by Vauxhall/Bedford's Research Dept in the 1950s, and particularly by the influence of General Motor's Maurice Olley.

I thank many people for their encouragement and input, and also the many pioneers in this area, particularly Richard Klein and (Professor) Josh Darling (Bath University UK). Also an academic at Sydney University, who (in 2002) advised me that 'deterministic chaos' was at the heart of caravan rollovers.

I also thank Scott Ross, Keith Berg and Daniel Weinstein for checking the pre-production draft of this book and Clayton's Towing for permission to reproduce their photographs for the front cover and Figure 4.1 - and also for their own practical insights in this matter.

All care has been taken to acknowledge the sources of the pictures used, do please advise if any need correcting etc.

I especially thank Colin Young, founder of the Caravan Council of Australia(https://www.caravancouncil.com.au), for his extraordinary dedication to making caravanning safer for all and his assistance over the years.

# DEDICATION

Maurice Olley was born in 1889, much at the time the first automobiles were being made. Following time his time as Chief Engineer of Rolls-Royce he moved to the USA in 1929 to work with the General Motors Research Division. He returned to the UK in the mid-1950s. I was immensely privileged to attend Olley's lectures during my own time at Vauxhall Motors' Research Laboratory.

Maurice Olley (1889-1972)

*"The entire history of mechanical engineering is of learning through failure. The prima donna type of mind is useless in engineering".*
Maurice Olley.

Maurice Olley was a quiet, humorous, unassuming and gentle engineer.

His work lives on in the 620-page Chassis Design: Principles and Analysis (http://www.millikenresearch.com/olley.html). The book was prepared from Olley's notes, some 27 years after his death, by Milliken and Milliken.

I owe a huge debt to Maurice Olley for encouraging my (lifelong) interest is this area. I dedicate this book to his name.

# ABOUT THE AUTHOR

Originally trained as an RAF ground radar engineer, Collyn Rivers spent a brief time with de Havilland designing power systems for guided missiles, before becoming a test engineer at the Vauxhall/Bedford Motors Research Test Centre.

He migrated to Australia in 1963, where he designed and built scientific measuring equipment. In 1971, Collyn Rivers founded what, by 1976, became the world's largest-circulation electronics publication, Electronics Today International.

From 1982 to 1990 he was technology editor of The Bulletin and also Australian Business magazines and in 1999 started two companies: Caravan and Motorhome Books, and Successful Solar Books (now rvbooks.com.au and solarbooks.com.au).

*Author, Collyn Rivers driving one of his (then) three Haflingers. These extraordinary vehicles were made in Austria during the 1970s - mostly for military use. At a mere 700 kg they can carry 750 kg, and are widely accepted as the most capable off-road vehicle yet made. Their 650 cc. twin cylinder motor is one half of a Porsche 356 engine. This one is ex-Australian Airborne Forces - designed to be dropped by parachute. Pic: Maarit Rivers.*

Collyn's books include the constantly revised Caravan & Motorhome Book, Caravan & Motorhome Electrics, Solar That Really

Works (for cabins, caravans and motorhomes) and Solar Success (for home and property systems).

Also available is his now globally-selling Disaster : Recovery (a guide to coping the critical few weeks after such events). It is published by Boiling Billy Books and is available in print form from all bookshops in Australia (many Dymocks branches have it in stock) and via email from booktopia.com.au. Its print version ISBN number is **978-1-925403-1-7**.

Amazon.com states: 'Disaster: Recovery is a down-to-earth book that explains how you and your family can cope during and particularly after disasters such as floods, bushfires and typhoons that can and do happen (not just may happen) around the world at any time.

'It is one thing to survive the initial impact of a disaster. It's quite another to survive the subsequent fallout. Water and electricity may fail, phones may stop working, roads may be impassable, food and fuel stores often run out of supplies. Medical help may be unobtainable as emergency services must scale priorities.

'This globally applicable book explains the whys and hows of surviving those vital first few weeks following a major disaster, and includes advice on the basics of being prepared. This is a book virtually every Australian home should have on the shelf.'

# PUBLISHING INFORMATION

**Publisher**: RV Books, 2 Scotts Rd, Mitchells Island, NSW, 2430. info@rvbooks.com.au

**Why Caravans Rollover & how to prevent it.**

**Copyright**: Collyn Rivers. All rights reserved. Apart from minor extracts for the purposes of review, no part of this publication may be reproduced, stored in a retrieval system, or transmitted in any form or by any means, electronic, mechanical, photocopying, recording, or otherwise, without the written permission of the publisher. Most drawings, graphs and tables in this book are copyright rvbooks.com.au.

**National Library of Australia - Cataloguing-in-Publication data**

Rivers, Collyn

Why Caravans Rollover & how to prevent it

**ISBN**: 978-0-6483190-6-1. Why Caravans Rollover & how to prevent it; Caravan Dynamics. I. Title.

**Production**: Text and artwork by Collyn Rivers.

**Publisher's Note:** This book is updated whenever deemed necessary. This is the first edition.

**Disclaimer**: Every effort has been made to ensure that the information in this publication is accurate. However no responsibility is accepted by the publisher for any error or omission or for any loss, damage, or injury suffered by anyone relying on the information or advice contained in this publication, or from any other cause. (The author would appreciate feedback relating to any errors and/or omissions).

**Front Cover Picture:** Reproduced by courtesy of Clayton's Towing.

# Table of Contents

| | |
|---|---|
| INTRODUCTION | 1 |
| **Part One** | 7 |
|   **CHAPTER ONE** | 8 |
|     Why and how caravans sway | 8 |
|     Dog-trailers | 8 |
|     Pig-trailers | 8 |
|     Hitch overhang | 9 |
|     Eliminating hitch overhang | 10 |
|     Fifth-wheeler - zero yaw | 10 |
|   **CHAPTER TWO** | 12 |
|     How tyres work | 12 |
|     How tyres are actually steered | 12 |
|     Slip angles | 13 |
|     Tyre wall stiffness and slip angle | 15 |
|     Tyre pressure and slip angle | 15 |
|     Laden tow-vehicle weight and caravan laden weight | 16 |
|     Tow vehicle weight matters | 17 |
|   **CHAPTER THREE** | 19 |
|     Understeer & oversteer | 19 |
|     Understeer | 19 |
|     Oversteer | 19 |
|     Understeer/oversteer, tyre pressures and slip angles | 20 |
|     Understeer/oversteer and mass distribution | 21 |
|     Dirt roads | 21 |
|     Summary | 22 |
| **CHAPTER FOUR** | 23 |
|   Further causes of instability | 23 |
|   Caravan length vs. weight | 23 |
|   Water-tank location | 24 |
|   Tow-hitch overhang and tow vehicle wheelbase | 24 |
|   Tow-hitch inserts | 25 |
|   Tow ball couplings | 25 |
|   Tow ball mass | 27 |
|   Towing capacity | 27 |
|   Gross Combination Mass | 27 |
|   Tow ball mass and weight distribution hitches | 28 |
|   Weight Distribution Hitch (WDH) | 29 |
|   WDH - pluses and minuses | 29 |
|   WDH adjustment | 30 |
|   Suspension modifications (tow vehicle) | 30 |
|   Airbag issues | 32 |
|   Independent suspension | 34 |
|   Caravan independent suspension | 34 |
|   Yaw (sway) reduction | 35 |
|   Friction yaw reduction | 36 |
|   Sprung dual-cam yaw reduction | 36 |
|   Electronic stability control - Al-KO ESC | 37 |
|   Dexter DSC | 38 |
|   Cruise control risk | 39 |
|   Side wind gusts from passing trucks (and when passing trucks) | 40 |

| | |
|---|---|
| Driver skill and reactions | 43 |
| Loading your rig | 43 |
| **Part one summary** | **45** |
| Tow vehicle and caravan behaviour | 45 |
| Towing stability - how does your rig rate? | 49 |
| Caravan Stability Questionnaire | 50 |
| A more technical explanation | 53 |
| **Part TWO** | **48** |
| **Part THREE** | **53** |
| CHAPTER FIVE | 54 |
| Suspension | 54 |
| Gyroscopic precession | 55 |
| Roll centre height - caravan beam axle suspension | 58 |
| Roll centre height - independent suspension | 60 |
| How much does roll centre height matter? | 61 |
| CHAPTER SIX | 63 |
| Critical speed | 63 |
| The single (fifth-wheeler) pendulum | 63 |
| Conventional caravan pendulums | 64 |
| Critical speed is rig and loading specific | 66 |
| Major stability factor | 66 |
| CHAPTER SEVEN | 67 |
| Designing a stable pig-trailer caravan | 67 |
| Suspension | 69 |
| Tow ball mass | 70 |
| Stability | 70 |
| **REFERENCES** | **78** |
| SAE International J8027 | 78 |
| Trailer and tow vehicle stability | 80 |
| Books | 81 |
| Papers | 82 |
| Tow Vehicle and Trailer Dynamics | 82 |
| Tyre Dynamics | 83 |
| Websites | 83 |
| **ACKNOWLEDGEMENT** | **84** |
| **>DEDICATION** | **85** |
| **>ABOUT THE AUTHOR** | **86** |
| **>PUBLISHING INFORMATION** | **88** |